Trust

Interdisciplinary Perspectives

Volume 4

This series presents interdisciplinary publications on trust. Trust is one of those rich concepts that have various meanings, some strong, others weak, some straightforward, others subtle. For instance, tryst, a medieval English word, means, in the context of a hare-hunt, the hunter who waits at the fringe of the forest to kill hares with a stick when they are driven out by big noises made by other hunters inside the forest. On the other hand, when one says, I trust that it will be fine tomorrow, its meaning is not strong.

More conceptually, trust is known as something that can be most useful in minimizing the costs of misunderstanding and transactions when the goal is to forge bridges and enhance bonds, launch joint undertakings, or establish reciprocal relationships. Without societies based on trust, sustained prosperity is far more difficult to achieve. The concept of trust is also very useful in understanding the propensity to take initiatives, to avoid risks, to enter into or withdraw from collaborations, and to shape and share values, norms and rules.

This series examines the meaning and its probable consequences of trust or distrust in various settings from various academic disciplines, not necessarily overtly confined to that of philosophy. Specialists of geography, anthropology, biology, neuroscience, sociology, political science and philosophy can venture to pen a volume from their respective angles. The resulting books will be both enlightening and enjoyable reading. This series welcomes proposals in these broad areas all over the world.

More information about this series at https://link.springer.com/bookseries/15199

Nobuhiro Kishigami

Food Sharing in Human Societies

Anthropological Perspectives

Springer

Nobuhiro Kishigami
National Museum of Ethnology
Suita, Osaka, Japan

ISSN 2509-7679 ISSN 2509-7903 (electronic)
Trust
Interdisciplinary Perspectives
ISBN 978-981-16-7809-7 ISBN 978-981-16-7810-3 (eBook)
https://doi.org/10.1007/978-981-16-7810-3

This Springer imprint is published by the registered company Springer Nature Singapore Pte Ltd.
The registered company address is: 152 Beach Road, #21-01/04 Gateway East, Singapore 189721, Singapore

Preface

An endlessly vast expanse of beautiful ocean and desolate tundra exists in the Arctic region of North America. It was in the summer of 1984, some 35 years ago, that I visited an Inuit community in Arctic Canada for the first time in my life for cultural anthropological research. Since then, I have studied Inuit culture in the Nunavik region and Montreal in Canada and Inupiat culture in Utqiaġvik, Alaska, USA. Almost all the persons I met there during my research as a foreign visitor were friendly people who welcomed me with smiles. It is no exaggeration to say that my unexpected encounter with the great nature and people of the Arctic fascinated me to continue my study of arctic anthropology.

As Japanese people are like Inuit and Inupiat in appearance, I could interact with them without any serious discomfort. On the other hand, the more I knew them, the more I got to know "mysterious" aspects about them that were beyond my comprehension. One of the aspects was their food sharing behavior.

One of the most impressive events I encountered in getting along with Inuit friends was that a hunter gave all or a part of his game, which he had caught with great difficulty, to other hunters or/and villagers without hesitation. As I had the idea of personal ownership deeply ingrained from Japanese thought, I did not understand his sharing behavior. Although I asked many hunters directly why they gave their game to other people, they always replied by asking why I asked them such a stupid question. As sharing among them was natural and the right thing to do, they did not have any particular or special reasons for sharing with other people in daily life. I had increasing difficulty to understand their sharing practices. This personal experience caused me to focus on food sharing as my research topic.

As I proceeded with my research, I learned that food sharing was almost universal in small-scale societies such as hunter–gatherer societies and that there was variation in the forms, ways, and effects or functions of food sharing practices in the societies. Further, people in small-scale societies tended to share not only food but also knowledge, tools, care, various activities, time, and space. Among them, food sharing was essential for humans to survive in a given environment in human history. In addition, I came to know that primates like the bonobo, chimpanzee, and orangutan share food with other individuals of the same species to some degree but do it in very different ways from human beings. Because of my personal experiences, I became interested in exploring why humans share food with others or give food to others. This book is a part of my research results concerning food sharing among hunter–gatherers in Arctic Canada and Alaska.

This book consists of five chapters. Chapter 1 reviews the perspectives and hypotheses of and approaches to food sharing practices of hunter–gatherers to propose problems that will be addressed in this book. While Chap. 2 concerns food sharing practices among Inuit in Canada, Chapter 3 deals with those among Inupiat in Alaska, USA. Then, Chap. 4 compares food sharing among three hunting–gathering societies including Inuit in Arctic North America, Aka in central Africa, and Aché in South America to examine the similarities and differences in food sharing practices among those groups. The last Chap. 5 discusses the food sharing practices of hunter–gatherers and summarizes the results.

This book is a result of 35 years of arctic anthropology research. The study and field research would not have been accomplished without the kind understanding and help from many people whom I cannot mention by name because they are too many in number. However, I must make explicit my sincere gratitude for a few. I thank the following persons who kindly helped me with my project in Akulivik, Utqiaġvik, and other arctic communities, and Montreal: Simon and Maggie Aliqu, Adamie and Amaly Anautak, Ainalik and Elisapie Alayco, Johanasie Qaqtuk, Geroge Barthe, Pita Atami, Jonny Epoo, Mary Palliser, Lizzie Palliser, Andrew Iqaluk, Simionie Weetalktuk, Victor Mesher, Annie Pisktie, Eugene Brower, Henry Brower Jr., Johnny Leavitte, Herman Ahsoak, Craig George, Hepa Takaliq, David Mohan, Barry Gunn, and Donat Savoie. I thank the following persons for their great understanding and support: Asahitaro Nishimura, Henry Stewart (Shunwa Honda), Kiyohiko Sakurai, Testuo Kikuchi, Kanji Masaoka, Hideaki Kiyama, and Yoichi Suzuki of Waseda University, Geroge Wenzel, Colin Scott, James Savelle, Bruce Trigger, Richard Salisbury, Fumiko Ikawa-Smith, Carmen Lambert, Toby Morantz, and Ludger Müller-Wille of McGill University, Molly Lee and David Koester of University of Alaska, Fairbanks, Kazuyuki Tanimoto, Tadasi Okudaira and Kaoru Sasaki in Hokkaido University of Education at Hakodate, Hiroaki Okada and Takashi Irimoto of Hokakido University, Minoru Ohshima of Otaru university of Commerce, Kazuyoshi Ohtsuka, Setsuko Ikuta and Marie Nakamura of the National Museum of Ethnology, Japan, Makoto Tanushi (artist), and my family including late Hatsuji (my grandfather), late Hisami (my grandmother), late Mitsugu (my father), late Teruko (my mother), Miwa (my wife), Mayo and Kahori (my daughters).

Finally, I would like to stress that the book is a result of the following projects: Arctic Challenge for Sustainability II (2019–2020), JSPS KAKENHI Grant Number JP19H00565 "A Comparative Study of Emergence, Current Status and Future of Indigenous Cultures in the Alaska and Northwest Coast Regions of North America" (2019–2021), JSPS KAKENHI Grant Number JP15H02617 "An Anthropological Study of Whaling Cultures in the Globalizing World: Conflicts between the Succession of Whaling Tradition and Anti-Whaling"(2015–2018), JSPS KAKENHI Grant Number 21401045 "Aboriginal Subsistence Whaling and Indigenous Rights in North America"(2009-2013), JSPS KAKENHI Grant Number

15251012 "Indigenous Distribution and Management of Marine Resources"(2003–2006), JSPS KAKENHI Grant Number 11691053 "Indigenous Use and Management of Marine Resources"(1999–2001), Minpaku Inter-University Research Project "A Comparative Study of Change, Current of Status, Future of Indigenous Societies along the North Pacific Rim: A Perspective of Human History" (2020–2021), and Minpaku Inter-University Research Project "Rethinking Gift Theory: An Interdisciplinary Comparative Study of 'Gift-giving', 'Exchange', 'Sharing'"(2012–2015). I greatly appreciate financial supports for these research projects from the Ministry of Education, Culture, Sports, Science and Technology—Japan, Japan Society for the Promotion of Science, and National Museum of Ethnology, Japan.

Osaka, Japan Nobuhiro Kishigami

Contents

1 Food Sharing in Hunting-Gathering Societies 1
1.1 Introduction 1
1.2 Definitions and Modes of Food Sharing 2
1.2.1 Definitions of Food Sharing 2
1.2.2 Types of Food Sharing Among Hunter-Gatherers 2
1.2.3 Food Sharing by Hunter-Gatherers and the Concepts of Exchange and Reciprocity 3
1.2.4 The Concepts of Exchange and Reciprocity, and Issues Therein 5
1.2.5 New Classifications of the Modes of Food Sharing 6
1.3 Social/Cultural Anthropological Studies 9
1.3.1 Hunter-Gatherer Societies and Food Sharing 9
1.3.2 Social Relations Centered Around Food Sharing 10
1.3.3 The Social Functions of Food Sharing 12
1.3.4 Psychological States Related to Food Sharing 12
1.3.5 Food Sharing and World Views 13
1.3.6 Food Sharing and Social Changes 14
1.4 Ecological Anthropological Research 15
1.4.1 Food Sharing as a Means of Ecological Adaptation 15
1.4.2 The Sharing of Lives and Food 16
1.5 Evolutionary Ecological Research 18
1.5.1 The Variance Reduction Hypothesis 18
1.5.2 The Kin Selection Hypothesis 19
1.5.3 Reciprocal Altruism 20
1.5.4 Tolerated Theft 20
1.5.5 Cooperative Acquisition 21
1.5.6 Food Sharing as an Adaptive Benefit 21
1.5.7 Sharing as Costly Signaling 22
1.6 Primate Research 22
1.7 New Directions in, and the Current Situation of, Food Sharing Studies 23

1.7.1 Social Network Analysis ... 23
1.7.2 The Need for Studies on Avoiding Sharing and Non-Sharing ... 26
1.7.3 Other Trends ... 27
1.7.4 The Current Situation of Food Sharing Research ... 27
1.8 Various Studies on Food Sharing Among the Inuit and Other Hunter-Gatherers in the Arctic Regions ... 29
1.8.1 Food Sharing in Alaska ... 29
1.8.2 Food Sharing in Arctic Canada ... 30
1.8.3 Food Sharing in Greenland ... 35
1.8.4 Similarities and Differences in Food Sharing in the Arctic Regions ... 37
1.9 The Potential Contribution of Studies of the Inuit, Yup'ik, and Iñupiat to Food Sharing Research ... 38
1.9.1 Differences in Food Sharing Between Small and Large Groups ... 38
1.9.2 Food Sharing and Family/Kinship ... 39
1.9.3 Diachronic Changes in Food Sharing ... 40
1.9.4 The Objectives of This Book ... 40

2 Food Sharing in the Inuit Society: A Case Study from Akulivik in Nunavik, Canada ... 41
2.1 Introduction ... 41
2.2 History of the Nunavik Region and Outline of the Akulivik Society and Economy ... 41
2.2.1 Recent Inuit History of the Nunavik Region in Canada ... 43
2.2.2 The Akulivik Economy and Society ... 44
2.3 Customary Food Sharing Practices in Akulivik, PQ, Canada ... 47
2.3.1 Outline of Customary Food Sharing Practices in the 1990s ... 47
2.3.2 Customary Food Sharing in Nunavik Society in the 1990s ... 48
2.3.3 Game Sharing Among Hunters ... 48
2.3.4 Food Sharing Through Meals ... 51
2.4 New Food Sharing: Food Distribution via the Hunter Support Program and Through Local FM Radio Announcement ... 53
2.4.1 Hunter Support Program ... 53
2.4.2 Enactment and Use of the Hunter Support Program in Akulivik ... 54
2.4.3 Effects of the Hunter Support Program on Akulivik Society ... 56
2.4.4 Impact of the Hunter Support Program on Hunters in Akulivik ... 60
2.4.5 Food Sharing Through Local FM Radio Announcement ... 61

2.5 Discussion 61
2.6 Conclusion 64

3 Food Sharing Among the Iñupiat in Utqiaġvik, Alaska, USA 67
3.1 Introduction 67
3.2 Food Sharing Practices in Traditional Iñupiat Society 68
3.2.1 Sharing Practices in Traditional Iñupiat Society 69
3.2.2 Meat Sharing After Hunting 69
3.2.3 Basic Characteristics of Food Sharing in Traditional Iñupiat Society 70
3.3 Whaling-Related Food Sharing 71
3.3.1 Whaling in Utqiaġvik 71
3.3.2 Butchering and Sharing of Whales by the Iñupiat in Utqiaġvik 72
3.3.3 Whaling and Standard Sharing of Whales in Utqiaġvik 73
3.3.4 Synchronic Variation in Whale Sharing in Utqiaġvik Village 75
3.3.5 Diachronic Changes in Sharing: A Comparison Between Two Generations 79
3.3.6 Secondary and Tertiary Sharing of Bowhead Whales 81
3.3.7 Sharing and Giving 89
3.4 Sharing of Food Other Than Whale Meat 94
3.4.1 Sharing of Game 94
3.4.2 Sharing at the Time of Butchering 95
3.4.3 One-Sided Sharing 95
3.4.4 Characteristics of Sharing 95
3.5 Conclusion 96

4 Comparing Food Sharing in Three Hunting and Gathering Societies 99
4.1 Introduction 99
4.2 Framework for Food Sharing Comparison 100
4.2.1 Forms of Food Sharing 100
4.2.2 Functions and Effects of Food Sharing 101
4.2.3 Characteristics of Food Sharing 101
4.3 Food Sharing in the Aka Group 102
4.3.1 Subsistence Activity and Society of the Aka People 102
4.3.2 Food Sharing of the Aka 103
4.3.3 Characteristics of Food Sharing of the Aka 105
4.4 Food Sharing in Aché Society 106
4.4.1 Subsistence Activity and Society of the Aché People 106
4.4.2 Food Sharing of the Aché 108
4.4.3 Characteristics of Food Sharing of the Aché 111
4.5 Food Sharing in the Inuit Society 112
4.5.1 Subsistence Activity and Society of the Inuit People 112
4.5.2 Food Sharing of the Inuit 113

4.5.3 Characteristics of Food Sharing of the Inuit 115
4.6 Comparative Analysis 116
4.6.1 Forms of Food Sharing 116
4.6.2 Functions and Effects of Food Sharing 117
4.6.3 Characteristics of Food Sharing in Hunter–Gatherer Groups 119
4.7 Conclusion 122

5 Food Sharing: Theoretical Considerations 125
5.1 What is Food Sharing in Hunter–Gatherer Societies like? 125
5.2 Hunter–Gatherer Sharing Form and Features 126
5.2.1 Food Sharing and Co-Presence 127
5.2.2 Food Sharing and Close Intimate Relationships 127
5.2.3 Food Sharing and Conversation 128
5.2.4 Food Sharing and Worldviews/Environmental Views 128
5.2.5 Food Sharing and Reciprocity 129
5.2.6 Food Sharing, Indebtedness, and Superior-Inferior Relationship 130
5.2.7 Food Sharing, Amounts of Games, and Population Size 130
5.3 Various Functions and Effects of Food Sharing Among Hunter–Gatherers 131
5.3.1 Environmental Adaptation 131
5.3.2 Economic Function 132
5.3.3 Political Function 132
5.3.4 Social Function 133
5.3.5 Cultural Function 133
5.3.6 Psychological Function 134
5.4 Sharing as a Human Characteristic 134
5.5 Conclusion 135

Brief CV of Nobuhiro Kishigami 137

References 147

Index 161

Chapter 1
Food Sharing in Hunting-Gathering Societies

1.1 Introduction

In hunter-gatherer societies, hunters frequently share their catch with other hunters or give it to others for no compensation. Such behaviors have been collectively described using the concept of food sharing, which has been considered a characteristic of hunter-gatherer societies. Many studies have been conducted on food sharing (e.g., Barnard 2002: 6–8; Damas 1972a; Dowling 1968; Service 1962, 1966: 14–21; Leacock and Lee 1982: 8; Lee 1999: 826; Lee and Devore 1968; Woodburn 1982).

The following researchers have reviewed studies on sharing: Price (1975), Kelly (1995), Kishigami (2002), Belk (2010), and Widlok (2017). They found that ways of sharing vary substantially between cultures; the origins and reasons for a certain type of sharing cannot be fully explained by looking at a single function or factor. In my earlier work, I summarized and examined diverse views on modes of food sharing. I then suggested that food sharing should be classified into several different types when it is analyzed. I criticized existing studies that perceive food sharing *en masse* as a form of reciprocity-based exchange. Based on this criticism, I proposed a new classification system for grouping, describing, and comparing food sharing practices (Kishigami 2003b, 2004a).

This chapter encapsulates the definitions and types of food sharing and reviews studies on food sharing among hunter-gatherers in sociology, anthropology, ecological anthropology, behavioral ecology, evolutionary ecology, primate research, and other disciplines. Thereafter, the chapter delves into findings on food sharing among the Inuit and other groups in Arctic North America and posits outcomes for food sharing studies. Through this process, the chapter suggests research questions that could help scholars who investigate hunter-gatherers in the Arctic and other regions to contribute to the study on food sharing.

N. Kishigami, *Food Sharing in Human Societies*, Trust 4,
https://doi.org/10.1007/978-981-16-7810-3_1

1.2 Definitions and Modes of Food Sharing

1.2.1 Definitions of Food Sharing

While food sharing is practiced most commonly among hunter-gatherers, it is also practiced in small farming and pastoral societies (Sahlins 1965: 187–225; Sillander 2019) where small groups of people often share food. Primates such as bonobos and chimpanzees do so as well (e.g., Silk 1987; Bercovitch 1988; de Waal 1989; Nishida 1992; Ihobe 1996; Muroyama 1999; Nishida and Hosaka 2001).

The behavior of food sharing in the animal kingdom is frequently described as food transfer, which can be divided into the following three forms: theft, tolerated theft, and gift-giving (donation). Primate researchers Nishida and Hosaka referred to the second and third forms of food transfer as food sharing and defined them as "the transfer of food from an individual to another without involving force" (Nishida and Hosaka 2001: 263–264). Their definition points to the fact that the owner of the food intends to share it. There are common features between food sharing among primates and Homo sapiens, such as sharing living space, shared experiences, and a superior memory. On the other hand, these species differ in their behavior of food sharing; for example, unlike modern humans, bonobos and chimpanzees do not exchange one object for another (Konma 2000: 12). Widlok (2017: 717), Belk (2010: XVII), and Hunt (2000) (among others) offered alternative definitions of sharing. This book employs the definition of food sharing put forward by Nishida and Hosaka since it can be used as a versatile basis for identifying phenomena. By adopting their definition, this chapter focuses solely on food sharing among hunter-gatherers in describing and analyzing such behavior.

1.2.2 Types of Food Sharing Among Hunter-Gatherers

Methods of classifying food sharing vary: Certain scholars have suggested that it should be categorized on the basis of the game's size (Graburn 1969; Altman and Peterson 1988), while others have recommended that it be based on whether it derives from production relations (e.g., Dahl 2000).

Altman and Peterson investigated a group of the Gunwinggu people who live in an outstation in West Arnhem Land in northern Australia. They found that Australian Aboriginal people have rules related to sharing and implied that it is important to differentiate large animals from small ones in examining the act of sharing (Altman and Peterson 1988: 76). While hunters' power over large game animals (such as kangaroos, crocodiles, and emus) is compromised by control from numerous parties, hunters have autonomy in how they deal with small animals (such as fish, lizards, and birds). Put differently, the former are shared as per the society's rules, whereas the latter are shared at the hunter's discretion (Altman and Peterson 1988: 92). In Inuit communities, large animals and food that can be supplied in abundance are shared widely, whereas small animals and limited amounts of catch are not (Graburn 1969: 66–67). Dahl observed food sharing among the Inuit in Greenland and proposed that the sharing of meat in accordance with rules be differentiated from the voluntary giving of meat. Dahl explained that the former type of sharing is grounded in production relations, whereas the latter is based on social relations (Dahl 2000: 175–177). Researchers in the African hunting-gathering societies studies group at Kyoto University perceive sharing in a chronological manner, using terms such as *primary*, *secondary*, and *tertiary* sharing, although these are not classifications (e.g., Ichikawa 1991; Tanno 1991; Imamura 1993, 1996; Kitanishi 1996, 1997, 1998, 2000, 2002).

There are multiple ways of classifying food sharing. Viewing it as a form of exchange based on the concept of reciprocity has been the mainstream in cultural anthropological studies of hunter-gatherer societies (ex. Service 1966). A majority of cultural anthropologists have been exploring food sharing among hunter-gatherers from perspectives of exchange and reciprocity since the following studies were published: Malinowski's study of the Kula exchange system (1922); the theory of gift-giving by Mauss (1924); the theory of exchange by Lévi-Strauss (1949), and the theory of exchange by Sahlins, which is based on findings from the above studies (Sahlins 1965, 1972). The subsequent section delves into the relationship between food sharing and the concepts of exchange and reciprocity.

1.2.3 Food Sharing by Hunter-Gatherers and the Concepts of Exchange and Reciprocity

Malinowski described seven gift-giving systems; these include *urigubu*, *sagari*, *wasi*, and *gimwali*, together with Kula, the ritual gift exchange system of the Trobriand Islands (Malinowski 1922). In the Kula exchange system, necklaces made of red shells (*soulava*) and bracelets made of white shells (*mwali*) are circulated among Kula participants from neighboring islands as though the items were relay batons: *soulava* is always passed in a clockwise direction, and *mwali* in a counter-clockwise direction. Malinowski viewed the Kula exchange system as an economic as well as a social and ritual one. He posited that such exchange was practiced against the backdrop of reciprocity, and pointed out that this system generated a sense of unity between different communities (which were geographically apart from each other) through the exchange of ritual gifts.

In investigating gift-giving, Mauss (1924) conducted case studies of the potlatch ritual (a gift-giving competition ritual) on the Northwest Coast of North America and the Kula ring in the Trobriand Islands. At a glance, gift-giving may appear to be a voluntary, spontaneous act. However, according to Mauss, obligations to "give," "receive," and "return" a gift are inherent to gift-giving. Therefore, this act leads to the reciprocal transfer of assets between the giver and the recipient. Mauss also carried out a case study of *hau* (spiritual power), which mediates the exchange of items in Māori communities in New Zealand, claiming that items given to others are regarded as possessing a spiritual power that leads the items to return to the original owner.

Lévi-Strauss greatly respected the fact that Mauss highlighted gift-giving and the resulting exchange (reciprocity). Notwithstanding, he objected to Mauss' explanation, which was based on the subjective views of indigenous people who believed that spiritual power is immanent in objects (cited in Godelier 2000). Lévi-Strauss maintained that exchange is the foundation of a society, and that living in a society is equivalent to exchange in terms of three dimensions: women, property, and language (Lévi-Strauss 1977: 145). He viewed marriage in the book "*Elementary Structure of Kinship*" as the exchange of women, and proposed the notions of restricted and generalized exchange. Restricted exchange occurred in parallel-cousin marriage. In this type of marriage, two groups exchange women in a reciprocal manner. Generalized exchange happened in cross-cousin marriage, where three or more groups participate and give women to other groups cyclically. Groups that have given women will receive women from different groups. Generalized exchange can integrate wider communities (compared to restricted exchange). Lévi-Strauss started his research with an understanding of social phenomena as exchange, and therefore perceived gift-giving as a form of exchange (Lévi-Strauss 1977: 145).

Lévi-Strauss (1949) influenced British anthropologist E. Leach and American anthropologist M. Sahlins; both elaborated new theories of exchange based on Lévi-Strauss's ideas. Leach believed that any form of exchange involves the exchange of the same item or different items. Based on this stance, he developed the symmetrical and asymmetrical exchange models (Leach 1982). His concept of asymmetrical exchange refers to scenes where items or services are given in exchange for other items or services such as items with monetary value, gratitude, and services. Leach asserted that in societies where asymmetrical exchange is practiced, eventually the flow of items and services exchanged between individuals will even out, while the value included in such exchanges involves intangible qualities, such as social status and prestige (Leach 1961).

Sahlins contended that in non-state societies without writing systems, reciprocity is a continuum of modes of exchange that include all kinds of exchange; he further noted that items, property, and social relations are interrelated (Sahlins 1965, 1984: 224–229). He defined reciprocity as the reversible movement of objects between two parties (Sahlins 1984: 225), and mentioned that generalized and negative reciprocity lie at both ends of the modes of exchange. Sahlins positioned balanced reciprocity between the two types of exchange: generalized reciprocity refers to altruistic exchange (i.e., generous exchange without expectations of anything in return) (Sahlins 1984: 233). In this type of reciprocity, items and property are usually given in a unidirectional manner, rather than a bidirectional one (Sahlins 1984: 234). Balanced reciprocity entails direct, obvious exchange. A typical example is an item with a value equivalent to that of the received gift is promptly given to the giver as a reciprocal gift (Sahlins 1984: 234). Negative reciprocity, at the opposite end of generalized reciprocity, refers to exchanges and attempts where one party tries to gain certain items or services from the other party without experiencing any losses (Sahlins 1984: 235).

Generalized reciprocity appears among members with a close kinship bond, while negative reciprocity emerges among those in an adversarial relationship. Balanced reciprocity appears among those in a relationship that can be positioned between kinship and adversarial ties (Sahlins 1984: 234). In other words, Sahlins identified the types of reciprocity that correspond with social distance/relationship.

Sahlins' model of reciprocity has been widely accepted in Western academic communities. Therefore, it is prevalent for food sharing in hunter-gatherer societies to be seen as reciprocity-based exchange that is voluntarily initiated by the giver (ex. Service 1966).

1.2.4 The Concepts of Exchange and Reciprocity, and Issues Therein

Exchange and reciprocity can be viewed either as the same act or different acts. Also, the meaning and use of reciprocity vary across researchers.

Lévi-Strauss (1977) and Sahlins (1984) used reciprocity and exchange synonymously. For them, reciprocity and exchange both indicate the bidirectional flow of items and property. Evolutionary ecologists employ the term reciprocity interchangeably with that of exchange, which signals a bidirectional flow (Hawkes 1992: 270–271). Certain economic anthropologists also use these concepts interchangeably. Meanwhile, other researchers clearly differentiate reciprocity from exchange.

Polanyi examined economic activity from the angles of reciprocity and exchange and put forward the notions of reciprocity, redistribution, and exchange as the integration principles of a social system in which economic processes take place. Polyani affirmed that a social system is required for such processes to occur (1957, 1975, 1977). Reciprocity refers to the transfer of property and services between different points in a symmetrical positional relationship, whereas redistribution involves the transfer of property and resources to a single center from which they will again be sent out (Polanyi 1977: 36). Exchange entails the transfer of property and services between two or random points within a system (Polanyi 1977: 36). Polanyi's definition of exchange can be considered equivalent to market exchange, as it relies on the existence of a market system (Polanyi 1977: 37). As per Polanyi's definition, exchange is characterized by the bidirectional transfer of property between two parties whereby both parties attempt to gain profit (Polanyi 1977: 42). Polanyi's definition of reciprocity is mostly similar to Lévi-Strauss' (1977) definition of reciprocity/exchange. However, Polanyi clearly differentiated reciprocity from exchange and gave them different meanings.

Hence, researchers have different definitions of exchange and reciprocity. What follows points to issues with the concept of reciprocity by Lévi-Strauss (1977) and Sahlins (1984). The current theory of exchange is centered around their ideas.

Both "generalized exchange" (by Lévi-Strauss) and "generalized reciprocity" (by Sahlins) are theoretical notions that indicate the unidirectional flow of objects such as women and food. In other words, these concepts do not anticipate the bidirectional flow of women and property between two parties. The former indicates circulation, whereas the latter implies giving from one party to another. Exchange signals the bidirectional flow of women and property between two parties, even if there are time differences between each transfer. I do not agree with using the concept of exchange or reciprocity in referring to the circulatory or unidirectional flow of objects.

Analysis based on reciprocity and exchange has been effective in studying the structures of cousin marriage and Japanese gift-giving (Lévi-Strauss 1977; Ito and Kurita 1984). In these areas of research, the following situations have been observed: certain groups exchange women, and certain groups or individuals exchange gifts; the groups and individuals involved have rules for exchanges and interactions. On the other hand, in hunter-gatherer societies, no groups or individuals that would engage in interactions for exchanging food and catch are clearly identified in a two-party relationship, except for cases in which specific partners identified by certain rules or close kin persons exchange such items. In hunter-gatherer societies, analysis using the concept of reciprocity may have limitations.

While reciprocity is a strong tool, it may also inhibit adequate comprehension. Konma stated, "when cutting out one single scene from the entire circle of exchange, it always appears to be unidirectional gift-giving. The power of this trick is stronger when it takes a long period of time for a reciprocal gift to be given" (Konma 2000: 44). Researchers with a deep knowledge of Lévi-Strauss' concept of reciprocity are likely to agree with Konma. However, I would still like to point out that no researchers have verified any propositions related to the above-mentioned "delayed reciprocity" by using empirical data.

Price distinguished sharing from reciprocity, maintaining that reciprocity involves an expectation that the giver of a gift will receive a reciprocal item with a value equivalent to that of the gift. Price pointed out that, in sharing, items and property flow in an asymmetric, direct manner, whereas in reciprocity they flow in a symmetric, direct way. Additionally, he noted that reciprocity entails multiple parties while sharing is practiced "within" a close relationship (Price 1975: 7).

Considering time differences, in the case of food sharing, it appears that food is constantly transferred (i.e., exchanged) between specific individuals—albeit the amount of food varies. Such situations can be viewed as a system in which the principle of delayed reciprocity is at play. In such situations, the practice of giving is habitually adopted, and the practice seems consistent with the model of delayed reciprocity. I believe it is necessary to start investigating such scenarios by exploring the practice of giving and receiving, rather than understanding food sharing as a form of exchange or reciprocity.

1.2.5 New Classifications of the Modes of Food Sharing

Starting in the early 1990s, the following studies were published: one that put forward the significance of demand sharing (Peterson 1993), and others that claimed sharing in hunter-gatherer societies is not exchange but redistribution (Woodburn 1998) or transfer (Hunt 2000). These investigations maintained that food sharing is not necessarily a form of reciprocity or exchange.

In my earlier work, I summarized studies by Peterson (1993), Woodburn (1989), and Hunt (2000), and showed that while food sharing had been dealt with *en masse* as a single category, it can be broadly classified into the following three modes: giving (giving or transferring), exchange, and redistribution. I also indicated that food sharing can be further divided into nine types when considering the following conditions: whether distribution is practiced in accordance with rules, voluntarily, or upon demand (Kishigami 2003b, 2004a).

I start with perceiving food sharing as a practice that can be objectively identified by a third party. In doing so, I draw on the three types previously proposed (giving, exchange, and redistribution) (Kishigami 2003b, 2004a).

Giving refers to the unidirectional transfer of objects from an individual (or group) to another. There is a tendency for such transfer to be described as gift-giving. However, since gift-giving in Japanese and other languages strongly implies an act of sending a special present or asset in special occasions, this book uses the term *giving*. Giving means the unidirectional transfer of food or an item from one party to another. In principle, the act of giving concludes at each transfer. Nevertheless, when giving is repeatedly practiced over a long period of time between several persons, food is transferred bidirectionally between them. Such a situation may result from practicing giving. I thus argue that it should not be seen as a form of exchange from the beginning.

Exchange refers to the bidirectional transfer of objects between individuals (or groups) and includes immediate or delayed exchange (in which a certain amount of time is taken for the subsequent transfer to occur) and asymmetrical exchange (in which different objects and services or gratitude are exchanged) (Leach 1985: 194–201). No research conducted for a period shorter than one year could collect sufficient data on delayed exchange, as studies on this type of exchange require many years of observation. In the case of asymmetrical exchange, it is difficult to gather solid evidence that proves that gratitude or reputation is given as a reciprocal return. Therefore, I only apply the category *exchange* to situations where it is clear that an individual who has received food is giving back food, an object, another recognizable service, or gratitude to the giver.

Redistribution entails circumstances where people other than the hunter collect catch from the hunter and give it to other people. K. Polanyi has already coined the anthropological concept of redistribution (Polanyi 1957, 1977). In other words, according to Polanyi, redistribution collectively refers to processes in which objects and food are gathered and sent to a center before being distributed (Polanyi 1957; 1977). Redistribution, which is used in this book, also, involves situations where an animal caught by a hunter is owned by people other than the hunter before the animal is distributed to other individuals.

To further divide giving, exchange, and redistribution into subcategories, the following three criteria can be considered: by rules, voluntary, and upon demand. It can be often difficult to distinguish between rule-based and voluntary giving. The latter is not subject to strict rules but affected by pressure from the community. As such, when giving is practiced in accordance with well-defined, socially shared rules, it is referred to as "rule-based giving." When giving is started by the initiative of, and demand from, the recipient, it is referred to as "giving upon demand" or "demand sharing" (Peterson 1993). "Voluntary giving" is defined as a situation where the giver's intention is respected. Gift-giving involves offering food and presents in the ritual of the potlatch on the Northwest Coast of North America. In this type of giving, the giver may forcibly give a gift. This can therefore be categorized as voluntary giving.

The terms *giving*, *exchange*, and *redistribution*, as used in this book, are meant for understanding the physical flow of food and objects. These concepts do not principally involve intangible factors, such as pressure from the community to engage in certain acts, or giving that is not considered reciprocal by the agent. I would like to establish that it is possible to apply the idea of reciprocity in understanding delayed exchange, asymmetrical exchange, and bidirectional transfer due to the cumulative practice of giving. Hence, while reciprocity can be used as an analytical concept, this book distinguishes it from descriptive notions that indicate the transfer of objects, services, and gratitude, such as giving and exchange.

When food sharing is categorized on the basis of the conditions that lead to sharing as well as the flow of food, nine modes (Table 1.1) can be identified. (To date, cultural anthropologists in English-speaking countries have viewed food sharing under a single category; they have not classified it into any types.)

The first type, "giving by rules" refers to situations where food is unidirectionally transferred from the hunter or food owner to other individuals in accordance with strict rules. In the second type, "voluntary giving," the hunter or food owner voluntarily gives food to others in a unidirectional manner. In the third type, "giving upon demand," the hunter or food owner gives food to others in a unidirectional manner upon demand. The fourth type, "exchange by rules," entails the exchange of food based on strict rules. The fifth type, "voluntary exchange," refers to the voluntary exchange of food. In the sixth type, "exchange upon demand," the exchange of food

Table 1.1 Classification of the modes of food sharing

	By rules	Voluntary	Upon demand
Giving A – – – – > B	Type 1	Type 2	Type 3
Exchange A ⇄ C	Type 4	Type 5	Type 6
Redistribution ↗ B A → D ↘ E	Type 7	Type 8	Type 9

A, C: the hunter or owner of meat; B, E: the recipient; C: distributors other than the hunter. The arrows indicate the directions in which food flows is initiated upon demand from the recipient. In the seventh type, "redistribution by rules," an animal caught by a hunter or food owner is distributed to others in accordance with strict rules by representatives and those other than the hunter or original food owner. In the eighth type, "voluntary redistribution," an animal caught by a hunter or food owner is voluntarily distributed to other individuals by representatives and those other than the hunter or food owner. The ninth type, "redistribution upon demand," involves the distribution of catch that is initiated upon demand from people other than the hunter, food owner, or representatives.

The types proposed above focus on the modes of practice. I have deemed them effective in describing the diversity and chronological changes of food sharing in certain hunter-gatherer societies, and in comparing different hunter-gatherer societies.

In this book, I propose to define food sharing as the transfer of food among individuals and groups or between individuals and groups in the modes outlined in Table 1.1. This differs from the traditional definition, which places food sharing under a single category.

The transfer of food practiced in any of the modes described in this section is referred to as food sharing. A plethora of research has been conducted on food sharing in hunter-gatherer societies. The subsequent section provides an overview of key studies on food sharing.

1.3 Social/Cultural Anthropological Studies

1.3.1 Hunter-Gatherer Societies and Food Sharing

Woodburn (1982) broadly categorized human societies into the following two types: those with the immediate-return system and those with the delayed-return system. In societies with the immediate-return system, people gather the harvest directly from labor. Obtained food will be consumed on the same day or within several days. It will not be stored or processed by complex methods. People in such societies are usually highly mobile and use simple hunting tools and everyday items that can be carried around and obtained easily (Woodburn 1982: 432). These people do not have a stratified society or class hierarchy. Meanwhile, in societies with the delayed-return system, a return on labor is not acquired immediately. In such societies, people invest in labor and store food in a scheduled manner. They tend to settle in a certain area and have a complex social organization.

Woodburn found that in hunter-gatherer societies with the immediate-return system, egalitarianism prevails in terms of wealth, power, and authority, and equality is maintained through various practices. The organization of such societies has four characteristics. First, the social group is highly flexible and its members are constantly changing. Second, individuals cohabit with others and decide (based on personal preferences) with whom to engage in hunting and gathering, trade and exchange, and rituals. Third, members do not depend on particular individuals in accessing diverse resources necessary for survival. Fourth, while sharing and interdependence are emphasized in interpersonal relationships in such societies, their members do not maintain any special long-term relationships or depend on certain individuals (Woodburn 1982: 434).

Woodburn affirmed that egalitarianism and sharing are intricately interrelated (Woodburn 1982: 440–442), and sharing plays an important role in preserving equal relationships between members of groups, as it can even out wealth and surpluses (Woodburn 1982: 442).

Barnard (2002) posited that societies that rely (or previously relied) on foraging, such as hunter-gatherer societies, be characterized by their mode of thought rather than their production. Barnard pointed out that the foraging mode of thought can be commonly observed in current and former hunter-gatherers. One of the inferences that can be derived from this mode is that hunter-gatherers consider saving to be an anti-social act and value sharing.

1.3.2 Social Relations Centered Around Food Sharing

Testart (1987) divided the sharing of catch in hunter-gatherer societies into the following two types: the non-Australian hunter-gatherer kind, and the Australian hunter-gatherer kind. Testart contended that the difference between the two types derives from differences in kinship systems.

Food sharing of the non-Australian hunter-gatherer type includes the form practiced by the Inuit, the San, and the Aka. In this type, the hunters who catch an animal initiate sharing (Testart 1987: 292). While differences in sharing have been observed between the Inuit and the San, a common feature between these groups is that there is a correlation between the frequency of sharing and the social and spatial distances in kinship that expand in a concentric manner, starting from the ego (Testart 1987: 295). In this kind of sharing, kinship does not affect how hunters own and use their catch. Testart referred to this type as System A.

The other form of sharing is seen among Australian hunter-gatherers. Testart called this type System B. In this type, the sharing of catch is initiated by those other than the hunter who catches it. Hunters are not allowed to own an animal after they catch it; the animal is collected and shared by people other than the hunter. Hunters are only allowed to keep the parts of the animal with the lowest quality, and are not permitted to share the other parts. A hunted animal is shared and received by the hunter's relatives and in-laws in the following order: immediate relatives, father-in-law, brothers-in-law, and brothers. The hunter, who owns the spear that first hit the animal, can only receive his share after all the others have received their (better) parts (Testart 1987: 292).

In Australian Aboriginal communities, members use two-section, four-section, and eight-section kinship systems (in which units are referred to as sections). Testart focused on the fact that Australian Aboriginal communities have organizations that comprise sections. A section forms mutually complementing relationships with other specific sections, such as relationships with blood relatives and in-laws, exogamous relationships between moieties, and relationships between different generations. The fact that a hunter who caught an animal cannot own or use it means that people in sections that have a mutually complementing relationship with the hunter's section can own and use the animal (Testart 1987: 296). In this way, System B constitutes the social structure of Australian Aboriginal communities. For example, an analogy has been observed between distribution and exogamy. Hunters may not have marital relations with women who belong to their own moiety (i.e., hunters cannot marry such women). In the same vein, hunters do not consume animals that they have hunted. Members of moieties that have a mutually complementing relationship with the hunter may consume the animal. The fundamental principle of this structure involves discontinuity from the closest party and linkage with the furthest party (Testart 1987: 299). Kinship is prioritized in Australian Aboriginal communities; it is not an exaggeration to say that kinship dominates their social and economic behavior.

Testart attempted to explain the difference between the non-Australian hunter-gatherer type and the Australian hunter-gatherer type from a Marxist perspective. He defined the social relations of production as relations that connect producers and non-producers that determine how products are divided between the parties (Testart 1987: 300). In sharing in the non-Australian hunter-gatherer type, a part of the hunted animal will return to the hunter at the end of the process. Meanwhile, sharing in the Australian hunter-gatherer type is practiced as though producers were completely deprived of their products for the sake of others. While producers cannot own or use their products, the products are collectively owned and used by the entire community. As shown above, hunter-gatherers of the Australian and non-Australian types have different social relations of production. Testart therefore considered that modes of production also differ between these types.

Apart from research by Testart, several other studies have focused on food sharing and social relations. In San communities, for instance, meat is reportedly shared on the basis of rules set in accordance with roles in hunting and kinship (Lee 1979, 1993; Marshall 1961, 1976: 297–298; Tanaka 1980: 95–96). In San communities (such as the !Kung, G|ui, and G||ana groups), close relatives are prioritized in the secondary sharing of meat (Marshall 1976: 297–298; Imamura 1993: 20). In Aka communities in the Lobaye region of the Central African Republic, the secondary sharing of meat is based on kinship (Bahuchet 1985: 371, 1990: 33). Tanno investigated multiple Aka Pygmy groups and claimed that Aka communities are by no means an aggregation of separate individuals: kinship dominates their sharing behavior and every aspect of their lives involves sharing (Tanno 1991). On the other hand, Kitanishi, who observed another Aka group residing near Linganga-Makaou Village, Dongou District, Likouala Department in the Democratic Republic of the Congo, reported that secondary sharing in the group was not necessarily based on kinship (Kitanishi 1997: 21).

In Alaskan Yup'ik and Iñupiat communities, as well as Canadian Inuit communities, food sharing functions as a part of the subsistence system, which is organized primarily around kinship ties. Hence, there is a strong correlation between food sharing and kinship. This connection has been frequently analyzed among hunter-gatherers in northern Siberia and Arctic North America (e.g., Burch 1988; Ellanna and Sherrod 1984; Kishigami 1995, 2000; Ready 2016; Wenzel 1981, 1995, 2000; Ziker and Schnegg 2005).

1.3.3 The Social Functions of Food Sharing

Food sharing is a social system consisting of several functions simultaneously. Ichikawa examined food sharing among the Mbuti and San peoples and found several functions, including a leveling function at the material level (Woodburn 1982), mitigating changes in the amount of catch and individual differences in hunting skills (Ichikawa 1991: 28). Food sharing has a function analogous to insurance in that it reduces risk and prevents malnutrition (Widlok 2017:47). Kent observed settlements of San hunter-gatherers in Kalahari and reported that food sharing could help maintain egalitarian relationships as well as generate and preserve a sense of unity among family and friends (Kent 1993). Studies have witnessed similar functions in societies such as the Yup'ik, Inuit, and Aka groups (e.g., Dowling 1968; Damas 1972a; Langdon and Worl 1981: 36, 39; Bahuchet 1990). Food sharing through communal feasts can produce and maintain identity (e.g., Kishigami 2013a, b, c). Wenzel, who investigated the Canadian Inuit, found that food sharing among the Inuit contributes to the well-being of the entire community rather than individual survival (Wenzel 1991).

1.3.4 Psychological States Related to Food Sharing

Certain researchers have discovered that skilled hunters share food to gain prestige in their community. In Australian Aboriginal communities, people know who the skilled hunters are (Altman and Peterson 1988: 80). Altman and Peterson asserted that skilled hunters always receive high prestige, irrespective of how they are treated and whether they can share meat. From such a perspective, hunters can gain prestige by giving their catch to others; this is one of the factors that leads to their food sharing.

Ichikawa revealed that food sharing has multifaceted significance because it may give rise to a sense of debt and unequal social relations (Ichikawa 1991). When food, property, or services are passed from one party to another, the recipient is expected to have a sense of debt, and the giver a sense of superiority.

In his work "The Gift" (1924), Mauss claimed that in societies without writing systems, while the recipient of a gift feels a sense of debt, the spirit of it always wants to go back to the original owner (giver). Thus, the recipient gives it back to the original giver after a certain period has passed. The philosopher Hitoshi Imamura clearly distinguished gift-giving from exchange (Imamura 2000: 5). He viewed the practice of gift-giving as a habitual or systematized behavior (Imamura 2000: 112). According to Imamura's definition, food sharing among hunter-gatherers is a type of gift-giving. Imamura emphasized a sense of debt when he discussed why humans voluntarily engage in unconditional gift-giving where they do not expect anything in return (Imamura 2000: 119). A sense of debt derives from feelings of gratitude and being in debt. Imamura stated that "a sense of debt is a basic composing factor of the existence of human beings" (Imamura 2000: 68). He affirmed that without a sense of debt, people cannot be genuine (Imamura 2000: 119).

Imamura divided gift-giving into competitive and non-competitive types. In competitive gift-giving, the parties involved reject sharing a sense of debt and compete for superior positions. In non-competitive gift-giving, the parties share a sense of debt through mutual gift-giving, thereby maintaining equality (Imamura 2000: 126). In the former type, a superior-inferior relationship is formed between the parties; in the latter type, the parties develop a collaborative relationship by sharing items.

Although Imamura's explanations are philosophical, they appear to correspond with ethnographic case studies of hunter-gatherers. However, much to the chagrin of this book's author, there is a dearth of anthropological research on human emotions. Hence, at present, it is difficult to conduct fieldwork that would provide accurate insight into changes in the feelings of people who are engaged in food sharing. Future studies should examine the emotions of the parties involved.

1.3.5 Food Sharing and World Views

Fienup-Riordan investigated the Yup'ik society on Nelson Island, Alaska. She found that symbolic cycles, such as one from animals to humans (and vice versa), as well as the ideology of exchange, form the foundation of all aspects of everyday life (including subsistence activity, kinship, marriage, the relationships between the hunting and the hunted, rituals, the sharing of seal meat, and gender-based division of labor) (Fienup-Riordan 1983). Fienup-Riordan affirmed that methods of sharing seal meat and the flow of seal meat embody cultural views of the world. In his case study of the Netsilik Inuit, Stewart (1991) also attempted to decipher perspectives of the world view and gender-based division of labor. In areas from northern Eurasia to northern North America, people widely share the perception that hunters should share the meat of animals they have hunted with others because the lives of hunted animals are given to humans (hunters and their wives) (e.g., Bodenhorn 1990, 2000; Kishigami 2000, 2005a, 2013a, b, c; Nuttall 1991; Omura 2021).

Bird-David (1990) conducted a case study of the Nayaka in India and focused on the idea (kinship metaphor) that the Nayaka consider the environment such as forests to be like their parents because it provides them with food. Nayaka people share food among them like siblings. Bird-David argued that the Nayaka's economy is based on the concept of "a giving environment" common among many hunter-gatherer societies; the Nayaka practice sharing upon demand while enshrining strict sharing ethics (Bird-David 1990).

These findings suggest that hunter-gatherers' ethics of sharing is due to their world views on the human relationship with the environment. However, although these ideologies help people socially accept and encourage certain behaviors, it is questionable whether they should be considered direct causes of behaviors like food sharing.

1.3.6 Food Sharing and Social Changes

Murphy and Steward (1956) predicted that if a small society with immature socio-economic strata were involved in the cash economy through trade, it would experience significant cultural changes, causing the ethnic culture (and the society that supports it) to break down. They noted this would lead the small society to be absorbed into the broader state system (Murphy and Steward 1956: 350, 353). Peterson and Matsuyama verified that hunter-gatherer societies respond to the penetration of the cash economy in a way that is much more complex and diverse than predicted by Murphy and Steward (Peterson and Matsuyama 1991). From the 1980s to the 1990s, sedentarization and the penetration of the cash economy were increasingly observed among many hunter-gatherer societies. Researchers noted that during this period, food sharing was conducted as an important socio-economic practice in hunter-gatherer societies across the globe, including the Inuit, San, and Mubti societies (e.g., Bodenhorn 2000; Ichikawa 1982; Fortier 2000, 2002; Kishigami 1995, 2000, 2004; Kitanishi 2000; Wenzel 1995; 2000).

The penetration of the cash economy as a part of economic globalization has had various impacts on food sharing. Matsuyama conducted research at the Gamardi outstation where the Australian Aboriginal Djinang people resided. He revealed that while the Djinang shared locally hunted wallabies and fish, they seldom shared food purchased from a market (such as flour and canned food) (Matsuyama 1994: 100). Hovelsrud-Broda carried out research in Isertoq settlement, Ammassalik Municipality, Greenland. She found that only country food was shared between different Inuit households in the village, and that cash or items purchased from a shop were seldom shared across households (Hovelsrud-Broda 1999: 45–46). However, market food is shared in other Inuit communities such as Akulivik Village, Nunavik, Canada, where I have carried out field research since 1984. There were different responses to the external forces on food sharing among several groups of hunter-gatherers.

It is critical to view changes in food sharing caused by sedentarization and the globalization of the economy as significant socio-economic shifts. However, little research has been conducted on the numerous effects of diachronic changes in the modes and functions of food sharing or the penetration of the cash economy on food sharing.

1.4 Ecological Anthropological Research

1.4.1 Food Sharing as a Means of Ecological Adaptation

Food sharing among hunter-gatherers is a means of closing the gap between successful and unsuccessful hunting; the amount of catch results from the difference in the individual level of hunting skills. Many ecological anthropologists have affirmed that sharing is equivalent to insurance for unstable, unpredictable changes in nature and can help protect individuals from scarcity (Gould 1982; Wiessner 1982; Cashdan 1985; Winterhalder 1986a, 1986b; Smith 1988). Wiessner (1982) claimed that the San people (specifically the !Kung group) mitigate the risk of food shortages within groups by sharing food between individuals and families. Additionally, they have a resource exchange system called *hxaro*. In *hxaro*, partners exchange resources while allowing other partners to use various resources including animals, water, and fuel in each other's area of residence (Wiessner 1982: 67–68; Barnard 2002: 8). In !Kung society, food sharing, together with the *hxaro* system, alleviates the instability of the food and other resources supply at the individual or family level.

Biesele and Howell examined !Kung society and reported that older members who had built and maintained trade and family networks for mutual aid received support and gifts through these networks once they could no longer look after themselves (Biesele and Howell 1981: 91). These findings suggest that food sharing during young adulthood can function as saving for retirement or insurance for life after retirement (Kelly 1995: 179).

Gould (1982) insisted that food sharing is practiced only in a highly unstable environment, where it generates social unity on which people can rely when necessary. He believed this to be the reason that the Ngangala people, who reside in a desert in Western Australia, widely practice food sharing. Gould also implied that people share food within a small, closed group when the environment is mild.

Kitanishi (1996) analyzed food sharing among the Aka people from an ecological perspective, shedding light on the following aspects: how the individual differences in the number of animals and honey harvested are related to socio-economic factors both within and outside the Aka culture; how such differences are rectified through food sharing; and how leveling the differences helps members of the camp to obtain sufficient food.

As mentioned above, certain researchers imply food sharing is a means of environmental adaptation among hunter-gatherers. The hypothesis central to these perspectives is referred to as the variance reduction model. This hypothesis will be discussed in the section below on evolutionary ecological research.

1.4.2 The Sharing of Lives and Food

Kaoru Imamura explored food sharing among the San people (specifically the G|ui and G||ana groups) in the central Kalahari Desert. Kitanishi observed the Aka people in the rainforest of Central Africa. Both researchers derived interesting results (Imamura 1993, 1996; Kitanishi 1996, 1997, 1998, 2000, 2002).

Japanese researchers' investigations of African hunter-gatherers have revealed that food sharing can be divided into several types. Imamura and Kitanishi, for example, examined food sharing by dividing the practice into the following levels: In *primary* sharing, hunters share the meat of the hunted animal at the site of hunting. In *secondary* sharing, meat brought back to the camp is shared among members of the camp. In *tertiary* sharing, cooked meat is shared among people at the camp (Imamura 1993; Kitanishi 1997). In addition, food is shared on the basis of rules in an obligatory manner in primary sharing, whereas it is voluntarily shared among people with close ties in secondary and tertiary sharing (Imamura 1993: 20–21; Kitanishi 1997). Moreover, the researchers observed a gender difference in food sharing behavior (Imamura 1993; Kitanishi 1997). There is a tendency for Japanese researchers, especially those educated at Kyoto University, to perceive food sharing as a product of face-to-face negotiation (Imamura 1993; Kitanishi 1997).

Imamura did not solely focus on food sharing as an independent topic; she aimed to elucidate the behavioral principles that consistently apply to collaborative work and sharing behavior. She looked at gathering and cooking (the main activities of San women) and reported that San women frequently shared food with people in proximity. Imamura pointed out that negotiations by San women were not dictated by a structure such as a dwelling relationship or kinship; they were conducted on the basis of interpersonal relationships formed through negotiation (Imamura 1993: 17–19). Imamura also indicated that it is crucial for San women to share the experience of eating food together. For food sharing and collaborative work (such as gathering plants), Imamura noticed that the San people voluntarily worked on tasks while sharing an overview of the entire effort. She suggested that such behavior was based on conformity, which refers to a sympathetic response evoked through interactions with others (Imamura 1993: 20).

Imamura referred to a system that controls various behaviors in face-to-face mutual negotiations on food sharing and collaborative work as the "sharing system" and claimed "they consider it essentially important to participate in all available opportunities and share experiences; therefore, collaboration and food sharing are collateral events that occur within the sharing system" (Imamura 1993: 21). Imamura also found that the sharing system functioned in units of camps when the San people were leading nomadic lives (Imamura 1993: 21).

Imamura did not explain food sharing from the angle of ecological adaptation; she stated that food sharing transcends practicality and is analogous to a type of ritual because it provides people with opportunities to "live again" by reflecting on the way of their living rather than opportunities to survive (Imamura 1993: 23). Moreover, Imamura pointed out that the San people practice food sharing not in an obligatory manner, but through face-to-face interactions, helping them rebuild their relationships with others (Imamura 1993: 23). Lastly, Imamura argued that conformity is the basis of San society, including food sharing (Imamura 1993: 23). In a separate investigation, Imamura asserted that the San people believe it is vital to share experiences, which includes food sharing. Sharing, which is central to San society, means sharing objects and experiences (Imamura 1996: 76–77). Imamura's view is almost equivalent to that of Tanno (1991) on food sharing among the Aka people as well as that of Ingold (1988: 283).

Kitanishi closely studied the relationship between food sharing and dwelling groups in the Aka community. Kitanishi discovered that except for the primary sharing of meat, the Aka people decide to whom they give meat not based on formal social ties (such as kinship), but on face-to-face relationships built within camp groups (Kitanishi 1997). Kitanishi found that the primary sharing of meat among the Aka was conducted by hunters in accordance with the degree of their contribution during the hunt. Obligatory strict rules applied to this level of sharing. Meanwhile, the secondary and tertiary sharing of meat was not obligatory but voluntary, where no strict rules were applied (Kitanishi 1997: 7–8, 21). There is a tendency for meat to be distributed widely throughout the camp, as it is repeatedly shared by people in the camp during secondary and tertiary sharing (Kitanishi 1997: 12).

Kitanishi made several interesting observations. First, when Aka women shared uncooked plants (such as fruit, yams, and cassava), they tended to give a large proportion to close relatives (Kitanishi 1997: 12–13). Second, the number of servings of cooked food largely depended on the amount of food, the size of the camp, prohibited food items, and the details of the food cooked (Kitanishi 1997: 14). Third, for the link between the size of the camp and food sharing among women, Kitanishi demonstrated that women naturally shared cooked food simply because they lived in the same camp (if the size of the camp was small) (Kitanishi 1997: 15). On the other hand, they were selective about with whom they would share food if the size of the camp was large. In the latter case, how cooked food would be shared was roughly determined on the basis of grouping in accordance with the positioning of huts. Kitanishi witnessed that those who had close kinship ties within the group frequently shared food with each other (Kitanishi 1997: 16). Fourth, food that the Aka strongly preferred was widely shared among many people (Kitanishi 1997: 22). Fifth, except for the primary sharing of meat, the owner of the food was not forced by any individual to share it with particular individuals. The owner of the food chose with whom he/she would share the food, considering the size of the camp, kinship, and spatial distance (Kitanishi 1997: 19). Kitanishi showed that in a large camp, the Aka chose with whom they would share food based on kinship ties. He affirmed that "it is important for people living in a forest camp to maintain an egalitarian network because they are isolated from other parts of the world, and constantly and repeatedly see a limited number of people. On the other hand, in a village camp, in which many people live, and which is more open to the external world, members of the camp need to be extremely selective in choosing a small number of sharing partners. Probably, kinship greatly affects their selection of such partners" (Kitanishi 1997: 22). Although Kitanishi did not analyze the differences in sharing and social relations between isolated groups and groups open to the external world, this proposition appears to help identify similarities in food sharing among other hunter-gatherer societies—including the Inuit after sedentarization in several settlements.

1.5 Evolutionary Ecological Research

Evolutionary ecological research has mainly investigated the behavior and evolution of living organisms. Starting in the late 1970s, certain anthropologists adopted evolutionary ecological approaches with the aim of elucidating how hunter-gatherers maintain hunting and gathering activities as well as systems of food sharing (Hawkes 1992; Kelly 1995; Kuchikura 2000; Smith and Winterhalder 1992; Winterhalder and Smith 1981).

Such research approaches are based on two assumptions. The first is that individual people (or individual organisms) will face trade-offs because time and resources are limited. Trade-offs refer to negotiations for several requirements that cannot be achieved simultaneously. If an individual makes a large investment for a certain purpose, the individual will inevitably have to reduce the size of other investments. This is an important cost–benefit question for an individual when he/she is deciding to take specific actions. Second, while individuals share conflicts of interest to a certain degree, it is extremely rare that they have the same conflicts of interest.

Since evolutionary ecology is grounded in such assumptions, several hypotheses have been developed that explain food sharing as an act of altruism. Hypotheses of evolutionary ecology and behavioral ecology for explaining food sharing among people include the following: the variance reduction hypothesis, the kin selection hypothesis, the reciprocal altruism hypothesis, the tolerated theft hypothesis, the cooperative acquisition hypothesis, the adaptive benefit or trade hypothesis, and the costly signaling hypothesis (Gurven 2004; Gurven and Kaplan 2002; Hawkes 1992: 274–276; Kaplan and Hill 1985a: 224–227; Smith et al. 2003; Tucker 2004; Ziker and Schnegg 2005).

1.5.1 The Variance Reduction Hypothesis

Achievements from hunting and gathering activities vary daily. At certain times, hunter-gatherers can obtain sufficient food, whereas at others they might not be able to obtain any food. Potential food shortages can be addressed by increasing diet breadth (diversifying the diet), storing, exchanging, or/and sharing food. Hunter-gatherers adopt such strategies by taking costs and benefits into account.

Winterhalder used a computer simulation in comparing food sharing and an increase in diet breadth from the perspective of costs and benefits. He concluded that to reduce the variance in everyday food resources without undermining efficiency in hunting and gathering, it is more effective to share food than to increase diet breadth

(Winterhalder 1986a, b). He argued that the following two factors determine the range of food sharing: environmental impacts on the difference in mean harvest rates among individuals, and the types of food resources harvested by members of the same camp/group. He also affirmed that there is a tendency for hunters who belong to the same camp to attempt to hunt different animals; therefore, people frequently share food when there is a large difference in harvest rates among hunters. Moreover, the effect of food sharing is maximized if the number of hunters is small (7–8 hunters) in a camp group (Winterhalder 1986a, b).

Smith asserted that even when the maximum amount of food is likely to be provided to the maximum number of people through food sharing, food storing is likely a more effective strategy than food sharing in the long run. However, Smith revealed that hunter-gatherers have continued to practice food sharing because they have a system that leads them to monitor each other's behavior. There is an expectation of reciprocity and they have a system of punishment for those who try to obtain food for free (Smith 1988: 240).

If individuals can no longer continue to maintain face-to-face interactions as the group grows, they will be able to use a strategy that allows them to unilaterally obtain food for free. For this reason, Smith and Boyd maintained that food sharing is not effective in a group with a large population (Smith and Boyd 1990).

1.5.2 The Kin Selection Hypothesis

From an evolutionary perspective, it is natural for parents to help their children and for individuals to help their relatives in the kinship network. This is because such behaviors help individuals pass on their genes to subsequent generations. Theoretically, individuals should share food with the relatives of the ego. A study on food sharing during meals in northern Siberia, Russia supports this hypothesis to a certain extent (Ziker and Schnegg 2005).

However, people share food with those other than the relatives of the ego. To address this contradiction, Hamilton coined the concept of inclusive fitness (Hamilton 1964), which refers to the degree of fitness not only of individuals, but also those within the kinship network. This means that even if the ego and the ego's children are sacrificed, they can still pass on their genes to subsequent generations (provided the relatives of the ego survive and pass on the same genes as those of the ego to future generations). Hence, people share food not only with those in their kinship group, but also with others who are external to the group. The Aché share plant food and food items purchased in a mission settlement within nuclear families. Parents pass these foods to their young children. Researchers support the kin selection hypothesis for the Aché. On the other hand, the Aché share meat and honey with those outside the nuclear family. In this sense, the Aché give and share food beyond kinship. Such behavior cannot be fully explained by the kin selection hypothesis. Kaplan and Hill asserted that in small groups such as those of the Aché, the range of food sharing can expand beyond close relatives if there are enough trade partners. Kaplan and Hill suggested that people in small groups prefer reciprocity to kin selection (Kaplan and Hill 1985a, b).

1.5.3 Reciprocal Altruism

Trivers (1971) proposed the idea of reciprocal altruism to explain altruistic behavior (such as food sharing) toward non-family members. Reciprocal altruism is used in answering the following question, which cannot be fully answered by the kin selection hypothesis: Why do people share food with non-family members? In other words, although individuals cannot increase their degree of fitness merely by independently engaging in altruistic behavior (i.e., giving food to others), they can increase their degree of fitness together with that of others if these individuals and others can repeatedly and mutually engage in altruistic behavior over a long period of time. This indicates long-term delayed reciprocity because the long-term benefits outweigh the short-term costs (Trivers 1971). Additionally, evolutionary ecologists use the concept of reciprocity in a way that differs from the use of cultural anthropologists (Hawkes 1992: 270–271). Sahlins (1965, 1984), for instance, employed the notion mainly when he referred to both the unidirectional and bidirectional flows of food and property. On the other hand, evolutionary ecologists utilize the concept verbatim (equivalent to "exchange") when they refer to bidirectional flows.

Male Aché hunters receive minimal rewards despite their high production and provision of food to a camp or residential group. Kaplan and Hill maintained that food sharing among the Aché people cannot be explained by the balanced reciprocity or variance reduction models (Kaplan and Hill 1985a, b). Moreover, this hypothesis has another issue: No ethnographic case studies have empirically validated delayed reciprocity (Pryor and Graburn 1980; Kaplan and Hill 1985a, b).

1.5.4 Tolerated Theft

If the failure to share resources leads to excessive costs, individuals may allow others to wield manipulation or force to give them the resource. When competitors see different levels of value in a resource, those who perceive higher value in the resource are likely to obtain it. This is because there is a trade-off relationship between the cost borne by individuals who are competing for the resource and the benefit that the owner of the resource will receive. Blurton Jones (1987) put forward the concept of tolerated theft, which explains that it is logical that the "haves" give the "have-nots" resources, since the have-nots can invest a higher level of competition cost in obtaining resources than the haves can. Blurton Jones's view is occasionally referred to as the tolerated scrounging hypothesis; it corresponds with a finding by Peterson (1993) that demand sharing has been widely observed among hunter-gatherer societies across the globe. However, while there are ostensible similarities between tolerated theft (Blurton

Jones 1987) and demand sharing (Peterson 1993), there is a fundamental difference between these two ideas. The former entails interactions between individuals who attempt to maximize benefits; the latter refers to food sharing practiced in accordance with specific social relations.

Food sharing among the Mikea people (hunter-gatherers living in southwest Madagascar) can be explained by this hypothesis (Tucker 2004). If this hypothesis is correct, conflicts over food should be visible; however, such conflicts were not observed in the case of the Aché (Kaplan and Hill 1985a, b). Moreover, if a large amount of food resource is available, tolerated theft can be expected to occur. However, in many hunter-gatherer societies, such as those of the Aka and the Inuit, food is primarily shared voluntarily or based on rules (e.g., Imamura 1993; Kitanishi 1997; Nuttall 1991; Kishigami 1995, 2000; Wenzel 1995, 2000). A plethora of cases cannot be explained using this hypothesis.

1.5.5 Cooperative Acquisition

This hypothesis explains the following: to increase the likelihood of catching an animal, hunters cooperate with each other while hunting, and the hunted animal is shared by those who participated in the cooperative hunting activity. In bowhead whale hunting by the Alaskan Iñupiat, those who participate in hunting, towing, and butchering of a bowhead whale will receive its meat. This case supports the cooperative acquisition hypothesis (Kishigami 2013a, b, c).

Nevertheless, this hypothesis cannot explain situations where Aché hunters share a hunted animal in the same manner, irrespective of whether the animal was hunted by a group of cooperating hunters or a single hunter (Kaplan and Hill 1985a, b).

1.5.6 Food Sharing as an Adaptive Benefit

Hill and Kaplan argued that Aché hunters who share food receive certain benefits that help to increase their degree of fitness, as existing hypotheses cannot fully explain food sharing among the Aché (Hill and Kaplan 1988a, b). Hill and Kaplan focused on the fact that skilled Aché hunters can have extramarital sex with the wife of another man as compensation for the meat they provided. They pointed out that, when including children from extramarital relationships, skilled hunters have more children than average hunters. They suggested that average hunters tacitly tolerate the situation whereby skilled hunters have extramarital sex with their wives to keep skilled hunters who can provide them with meat in the camp, so that members of the camp can receive more meat (Hill and Kaplan 1988a, b). In the above case of sharing in the Aché community, the parties involved exchange different types of value for mutual benefit. Hypotheses that explain such a case are referred to as mutualism or trade hypotheses. Notwithstanding, while the exchange of meat and sex may be appropriate in Amazonian culture in general, such exchanges are not necessarily common in other hunter-gatherer societies (Kelly 1995: 178).

1.5.7 Sharing as Costly Signaling

Hunters can let others know of their superior hunting skills by showing them off and giving others the largest amount of meat possible. In this way, skilled hunters can increase their likelihood of finding favorable hunting partners and obtaining more sexual partners. Through such a practice, hunters not only survive but also have many children (Gurven 2004; Smith, Bird, and Bird 2003). The earlier explanation is referred to as the costly signaling hypothesis or the show-off hypothesis.

The Dobe Ju/'hoansi, a group of the African San people, complain about, or ridicule, successful hunters when they receive meat from the hunters by saying things like, "It's a small animal" and "I only received a small share." In this sense, they have developed a cultural system that does not allow (although ostensibly does permit) specific hunters to build a reputation or prestige (Lee 1993). On the other hand, in other hunter-gatherer societies, skilled hunters have received prestige by sharing their catch (Widlok 2017: 50). In certain hunter-gatherer societies, food sharing is used to gain a reputation in the community or political status (e.g., Kishigami 2000; Ready 2016). However, the costly signaling hypothesis on its own cannot fully explain all known types of food sharing.

1.6 Primate Research

Food sharing is not limited to humans. New World monkeys (such as marmosets and tamarins) and great apes (such as chimpanzees and bonobos) also share food (Silk 1987; Jaeggi and Gurven 2013). Unlike humans, these animals do not proactively give out food (Muroyama 1999: 149).

In the case of chimpanzees and bonobos, food sharing occurs most frequently between mothers and children aged up to around six years (Ihobe 1996: 25–26). Adult male chimpanzees mostly share meat; bonobos mostly share plant food (Ihobe 1996: 25–26).

In the case of chimpanzees, high-ranking males own meat. Such male chimpanzees give meat to older ones, and to male friends as well as females in heat (Ihobe 1996: 25–26). Although adult male chimpanzees proactively share meat with other close males, they do not share with those with whom they do not have peer relations. Adult male chimpanzees use food sharing as a political tool that helps them improve their social status, enhance unity, and negotiate social relations (Bercovitch 1988; Nishida 1992; Ihobe 1996: 26). A characteristic among bonobos is that they may share food in exchange for mating opportunities (Ihobe 1996; 25–26). According to de Waal, chimpanzees exchange food for certain behaviors such as grooming (de Waal 1989). Matsuzawa pointed out that food sharing helps to generate a sense of closeness among chimpanzees and the act of food sharing is a joyful experience (Matsuzawa 2018: 183, 203–204).

Ihobe examined food sharing behavior among chimpanzees from an evolutionary perspective and claimed that food sharing among adult male chimpanzees developed as a political negotiation tool used to adjust social relations (Ihobe 1996: 25–26). Moreover, Ihobe revealed that food sharing among people is highly diversified and asserted that the practice "developed for the purpose of adjusting social relations between individuals, as well as building and maintaining collaborative relationships between groups, rather than to ensure a food supply" (Ihobe 1996: 25–26).

A comparative study discovered that among 69 primate species, 30 do not share food at all, 39 do so only with their offspring, and 15 do so between adult animals (Jaeggi and Gurven 2013: 188). Bonobos and chimpanzees, for example, engage in behavior that is considered food sharing. Therefore, these animals and humans may have similarities: for example, sharing lives and experiences, as well as having a superior memory in highly social, small groups. However, there is a significant difference between bonobos and humans, as bonobos do not exchange one object for another (Konma 1999: 12). Hence, food sharing is a behavior unique to humans (Widlok 2017: 32). Diversity in food sharing behavior among humans needs to be explained by drawing on cultural, social, and historical factors that are specific to humans, in addition to biological factors (such as language ability).

1.7 New Directions in, and the Current Situation of, Food Sharing Studies

1.7.1 Social Network Analysis

Toward the end of the 1990s, an increasing number of scholars who study food sharing conducted social network analysis. Previously, the analysis was performed in disciplines such as function theory, cultural theory, behavioral ecological studies, and sociology. One of the early researchers who incorporated the social network approach in resource sharing research was Wiessner, who investigated the *hxaro* system in the !Kung group in Africa (Wiessner 1986, 1998). Thereafter, the social network approach was adopted for studies on food sharing among indigenous people in northern Siberia, Arctic Canada, Alaska, and Indonesia (Burn Silver et al. 2016; Collings 2011; Collings et al. 2016; Dombrowski et al. 2013; Dyble et al. 2016; Kofinas et al. 2016; Magdanz et al. 2002; Natcher 2019; Nolin 2010, 2012; Ready 2016; Ready and Power 2018; Ziker 2007; Ziker and Fulk 2018; Ziker and Schnegg 2005). The approach statistically analyzes, in units of households, the frequency and methods of sharing food between households, as well as the types of networks used for food sharing, with the aim of identifying the nature and characteristics of networks between households. Such research also attempts to link existing social anthropological and behavioral ecological approaches. Below are a few representative studies.

Ziker (2007) scrutinized food sharing among the Dolgan and Nganasan peoples in the Taymyr region in northern Siberia, Russia from the perspective of a social network. Ziker pointed out that during the post-socialism period, subsistence activities and food sharing in the wider community had social significance for indigenous people in Siberia and helped them maintain nutritional well-being. Ziker and Schnegg found that households in the same family network in these communities were sharing meals in a reciprocal manner (Ziker and Schnegg 2005). Moreover, Ziker and Fulk looked at sharing networks among women and revealed a relationship between the sharing of country food and the following aspects: kinship between households, reciprocal sharing, the impacts of interaction, economic necessity, and social connection (Ziker and Fulk 2018).

Nolin observed the social network of food sharing among people in Lamalera Village, Indonesia who make a living from fishing and whale hunting. Nolin found that those in Lamalera frequently practiced reciprocal sharing among households with close kinship ties that were located close to each other. Nolin stated that reciprocal altruism is a key driver for people in Lamalera to practice food sharing and they select sharing partners primarily based on kinship ties and the distance between dwellings (Nolin 2010). High-status households gave and received more than other households. This finding corresponds to reciprocal altruism. Nolin also demonstrated that leaders were giving more, since such an act entails sharing-as-signaling by high-status households (Nolin 2012).

Collings and his colleagues investigated the Inuit in Ulukhaktok in Canada's Northwest Territories and discovered that food sharing plays a vital role in integrating the community (Collings et al. 1998). Collings analyzed the network of food sharing and discerned the difference between hunters and wage earners, explaining that while the former was building connections with extended relatives and kin, the latter focused on connections with parents and siblings (Collings 2011). A 2009 research of the social network of country food exchange in 122 households provided several fascinating findings. Collings and colleagues found that exchanging food is important not only from a nutritional perspective but also from political and economic angles. Members of each extended family eat together at least once a day; those who give others food are more respected than those who do not (Collings et al. 2016: 35–36). In addition, they found a strong ethic to share food with the elderly. Collings and colleagues recently revealed that in the village that they observed, the number of households in which a single man or woman is the head of household has increased to nearly 50% of the village's total households. They pointed out that such households are weakly associated with the village's food sharing network, implying that they have limited opportunities to obtain country food and may experience food security issues (Collings et al. 2016).

From July 2013 to July 2014, Ready (2016, 2019) carried out fieldwork in the Inuit village of Kangiqsujuaq in Nunavik, which faces Canada's Hudson Strait. She examined the sharing of catch between households as a household subsistence strategy in addition to the village's social structure. She scrutinized the relationship between data collected from 110 households in the village and its social network. The village has a population of approximately 800. The subsequent paragraph explains her results.

In Kangiqsujuaq, members of the community almost always share food with parents, children, siblings, other family members, and relatives of the ego. Food sharing in the village is closely associated with kinship. Food sharing among close relatives is characterized by reciprocal sharing and mutuality (Ready 2016: 37). Households with stable cash incomes and hunters are more proactive in engaging in hunting and fishing and sharing their products with other households. Additionally, hunters in these households reciprocally share their catch with households that have a similar level of income and are involved in hunting and fishing. These hunters share their catch with households with a female head and those that contain older persons (Ready 2016: 113). Meanwhile, low-income households that do not frequently engage in hunting or fishing do not necessarily receive food from affluent households. There is a tendency for low-income households to share food with those in similar situations (i.e., low-income households that do not frequently engage in hunting or fishing) (Ready 2016: 113–114). Ready discovered that since every household in the village does not share food (including catch) with all other households (including those in need), food sharing of this type does not necessarily benefit the entire community. Given this, Ready argued that food sharing in the village is not an effective means of adapting to climate change for low-income households with low productivity in hunting and fishing (Ready 2016: 117). Instead, food sharing maintains inequality between households in the village. Since generosity is the source of power in Inuit society, affluent households maintain political influence and power through the practice of sharing (Ready 2016: V, 106, 118). Ready affirmed that, although wealth is redistributed through food sharing among the Inuit, the act of redistribution strengthens the village's social structure. She asserted that this type of food sharing thus creates inequality and is used to justify such a structure. Although food sharing has long been considered a way to maximize a community's well-being (Council of Canadian Academies 2014: 78), Ready pointed out that low-income households with low productivity in hunting and fishing can contribute to the community's well-being to a limited extent. Ready's research is novel in that she illustrated that the current form of food sharing among the Inuit is linked to inequality and a power gap in the community. One issue with her research is that she did not examine food sharing through daily meals. Had Ready explored this aspect, she would have shed light on how extended families are likely to behave when sharing food (Kishigami 2018).

A research team from the University of Alaska, Fairbanks led by Kofinas investigated the mixed economy of Wainwright and Kaktovik, Alaska from 2009 to 2010, scrutinizing the social network among households. These two communities are characterized by the group hunting of bowhead whales as well as sharing and feasts at a community level. By considering changes over the past 30 years, the research team concluded that the mixed economy is not a temporary practice in these communities, as households there are still deeply involved in the following elements of mixed economies: a market exchange, subsistence activities, and traditional social relations. Remarkably, the two indigenous Alaskan communities and the Canadian Inuit have differences in food sharing. The research team discovered that high-income households tend to be more productive in hunting and provide other households with food, both in the two Alaskan communities and among the Canadian Inuit. However, the former differs from the latter in two aspects. First, unlike the case of the Labrador Inuit (Dombrowski et al. 2013), in the two Alaskan communities examined, low-income households with a good harvest provide a certain amount of food to other households. Second, in contrast to the Nunavik Inuit (Ready 2016), low-income households with poor harvests in Wainwright and Kaktovik receive a significantly larger amount of food by weight compared to other groups. These differences between Alaska and Arctic Canada may derive from the recognition and social significance of the above-mentioned group hunting of bowhead whales, one of their primary subsistence activities.

As explained in this section, studies involving social network analysis have uncovered a tendency for households of hunter-gatherers and hunter-fishermen to mutually share food with neighboring households of close relatives. Further, the more a household gives, the more respected it is. Such a household thus gains strong social influence and political power. As such, it is highly likely that food sharing is linked to leadership and political power in the community. Moreover, in Arctic North America, while high-income households with high productivity in hunting and fishing tend to give a large amount of food to other households, these affluent households also strongly tend to engage in reciprocal food sharing among themselves.

As described above, recent years have seen an increase in the number of studies involving social network analysis with the aim of shedding light on the relationships between actors and among households in the context of food sharing. Studies on the Canadian Inuit and Alaskan indigenous people have greatly contributed to this area of research. Social network analysis is also an attempt to integrate evolutionary ecological and social anthropological approaches.

1.7.2 The Need for Studies on Avoiding Sharing and Non-Sharing

Some of the latest studies have indicated that research on food sharing should not be limited to sharing but should also include non-sharing. The Mikea people in Madagascar (Tucker 2004, 2019) and the Mpimbwe people in Tanzania (Kasper and Borgerhoff Mulder 2015), both hunter-gatherers, rarely share food with people outside their immediate families and tend to avoid food sharing, although they view food sharing as a social ethic. Berman (2020) conducted fieldwork in the small village of Jajikon in the Republic of the Marshall Islands (RMI) from 2009 to 2013. The village has a population of 250; its members enshrine generosity and giving as their shared culture. Through her fieldwork, Berman found that while people in the village practice sharing, they also avoid it in everyday social and economic activities. She discovered that avoiding sharing plays a critical role in creating and preserving social relations in the village. People in the village avoid sharing without undermining existing social relations by co-constructing conversations with their relatives and neighbors. Berman advocated for economic and Oceanian anthropology to explore the avoidance of sharing as well as the practice of sharing in everyday life. Please note that Berman's idea of sharing refers to one-sided transactions with no expectation of a return (Berman 2020: 219).

1.7.3 Other Trends

In addition to the trends mentioned above, two more have been witnessed in recent research on food sharing. First, several studies have focused on women's role in food sharing. There was once a dearth of research on women's role in food sharing among hunter-gatherers. Quintal-Marineau conducted a study on Inuit women's role in modern subsistence activity in Kangiqtugaapik (Clyde River), Nunavut Territory, Canada. In the current Inuit community, in order for the men to engage in hunting and fishing, they must purchase items such as rifles, snowmobiles, canoes, outboard motors, gasoline and oil, and bullets. Inuit men can continue hunting and fishing because women contribute part of their incomes (from paid labor in the village) to the hunting and fishing activities of their sons, husbands, and brothers. Inuit men share the products of hunting and fishing with the women. Quintal-Marineau reported that in the modern Inuit community, women play a certain role in hunting and fishing activities as well as sharing the products of such activities (Quintal-Marineau 2016, 2017, 2019; Quintal-Marineau and Wenzel 2019). In this sense, there has been a rise in the number of studies on women's role in food sharing.

Second, recent studies tend to explain the origins of food sharing and reasons for its practice not by pointing to one factor but instead combining multiple factors (Kelly 1995: 180–181). Evolutionary ecological studies have already made this shift. Tucker (2019) claimed that food sharing should be explained as a practice that involves a complex web of history, strategies, and social systems.

1.7.4 The Current Situation of Food Sharing Research

Food sharing among hunter-gatherers is a social practice that allows people to distribute and access food. It has been beneficial at the individual, group, and societal levels. Most existing research on food sharing has aimed to clarify the origins of, and reasons for, food sharing among hunter-gatherers (or individual apes). Much of the research is based on explanations that food sharing is for ecological adaptation, and has social and economic functions and effects; most offer explanations from an evolutionary ecological angle. On the other hand, cultural and social anthropologists have been providing cultural explanations that center on correspondence between world views, social structures, and activities of daily life.

The ecological adaptation hypothesis focuses on food sharing as a means of reducing instability in food acquisition due to the nature of hunting and gathering activities. This hypothesis considers food sharing to be a type of insurance that mitigates the risk of failure in hunting at the individual level, closing the gap in hunting skills and amount of catch. Many researchers support the ecological adaptation hypothesis (Kelly 1995).

Several Hypotheses emphasize the socio-economic functions and effects of food sharing. In many hunter-gatherer societies (especially those with an immediate return system), food sharing has material and political leveling functions and helps to generate, maintain, and integrate social relations. Provided that food sharing has such effects and functions, the question remains: Can such hypotheses provide effective explanations as to why hunter-gatherers practice food sharing?

Evolutionary ecological hypotheses primarily provide explanations based on organisms' altruistic behavior. Major examples of these hypotheses include the kin selection hypothesis, the reciprocal altruism hypothesis, the tolerated theft hypothesis, the cooperative acquisition hypothesis, and the costly signaling hypothesis. Case studies of the Aché, among others, have found that these hypotheses are not necessarily correct (e.g., Kaplan and Hill 1985). Marlowe (2010) examined food sharing practices of the Hadza hunter-gatherers and argues for six relevant evolutionary ecological hypotheses concerning food sharing: (1) nepotism (food for kin), (2) mate provisioning (food for mating access or bond maintenance), (3) not-in-kind exchange (food A for food B), (4) in-kind-delayed-reciprocity (food A now for food A later), (5) costly signaling (food to advertise ability and receive benefits), and (6) tolerated scrounging (food for peace). Marlowe notes, "tolerated scrounging seems most apt to explain the widespread sharing of meat and honey when they cannot be hidden" among the Hadza hunter-gatherers (Marlowe 2010: 251).

However, many cultural and social anthropologists think that these hypotheses do not convincingly explain many food sharing practices conducted in the daily life of hunter-gatherer societies. Widlok insists that as evolutionary ecological theories of sharing have focused on the beneficial outcomes of sharing at societal and individual levels, they have not paid attention to process. I believe that the social, cultural, and historical contexts in which food sharing occurs should be taken into consideration.

According to a few studies, hunter-gatherers practice food sharing because they associate it with perceptions of the world that symbolize the relationship between people and the environment, as well as an ethical perspective that emphasizes generosity. Although such studies positively view the act of food sharing and explore its reasons by drawing on recommended ideologies, an ideology per se does not lead to any act. Hence, I do not believe that these studies provide effective explanations.

Studies that focus on social relations are useful in understanding how food sharing is practiced. Notwithstanding, the reasons people share food with non-relatives cannot be fully explained merely by pointing to relationships (such as family ties).

While primates also share food, a significant difference between primates and humans is that food sharing among humans is markedly characterized by exchange and reciprocity. Hence, exchange and reciprocity may be attributes that indicate humanity.

In recent years, certain studies have analyzed food sharing from the angle of the social network, whereas others have examined avoiding sharing, as well as women's role in sharing. The following standpoint has been gradually adopted as a mainstream view: the practices of specific types of food sharing should be explained not by relying on a single factor but by drawing on multiple factors or cumulative factors. This review found a lack of research on the psychology of the actors involved in food sharing (Kameda et al. 2002) as well as historical changes in food sharing.

1.8 Various Studies on Food Sharing Among the Inuit and Other Hunter-Gatherers in the Arctic Regions

Food sharing among the Inuit was seldom discussed as a main research topic or compared before the Eighth International Conference on Hunting and Gathering Societies in 1998 at the National Museum of Ethnology in Osaka, Japan. The results of one session during the conference helped researchers compare food sharing in modern Inuit society in areas ranging from Alaska to Greenland (Wenzel, Hovelsrud-Broda, and Kishigami eds. 2000). This section discusses key investigations on food sharing in Inuit communities and the outcomes of these studies. In doing so, the section compares three areas: Alaska, Canada, and Greenland.

1.8.1 Food Sharing in Alaska

Studies on food sharing and exchange among the Iñupiat and Yupiit in Alaska include those by Burch (1970, 1988), Spencer (1959: 167ff.), Bordenhorn (2000, 2005), and Fienup-Riordan (1983). Langdon and Worl also summarized food sharing and exchange in Alaska (Langdon and Worl 1981).

Burch reconstructed the history of the Iñupiat society in northwest Alaska in the early nineteenth century (Burch 1980, 1998). He found that the Iñupiat in the area have engaged in various forms of interaction, including giving food and other resources (e.g., *aviktuuzaaq*, *nigiq*, *pikszak*, and *pigziaq*), direct exchange (*simmiq*), trading between partners (*niuviq*), and theft (*tiglik*) (Burch 1988). He pointed out that generalized reciprocity (i.e., giving without expecting reciprocity) is only practiced within each local family, which is a nomadic group comprising a single extended family (Burch 1988: 109–109).

In the Iñupiat community of Utqiaġvik (Barrow), Alaska, people currently share not only food but also tools, knowledge, childcare, time, space, and cash. Bodenhorn stressed that food sharing is a practical process based on social relations through which food is permanently flown from one individual to another (Bodenhorn 2000:

44). In most cases, the Iñupiat reciprocally share food among relatives. Contrarily, studies suggest that they also engage in unidirectional giving, where they do not expect any reciprocity. Seals and caribou tend to be shared only among hunters and relatives. Meanwhile, large animals such as beluga whales, polar bears, and bowhead whales are shared among all members of the village (Bordenhorn 1990: 61). Additionally, people invite others to meals and give food to the elderly, single people, orphans, widows, and those in need (Bordenhorn 2000: 43–44). Since the Iñupiat practice unidirectional giving without expecting reciprocity irrespective of kinship ties, Bordenhorn affirmed that such a practice should be explained by referring to factors other than social ones. According to Bordenhorn, the Iñupiat hold the following view: animals come to hunters with righteous hearts to be hunted as long as the hunters do not monopolize the harvest but share it with others; the more meat a hunter shares, the more animals come to him (Bordenhorn 2000: 44, 47). Bordenhorn revealed that food sharing has always been an important practice in the Iñupiat community because its members share food on the basis of the above-mentioned world view (Bordenhorn 2000). She also argued that in the modern socio-economic context, food sharing is a way of adapting to modern society because Iñupiat people who do not have country food (e.g., the meat and/or other parts of bowhead whales)—which cannot be purchased with cash—rely on sharing food to obtain it (Bordenhorn 2000). In this manner, Bordenhorn attempted to explain food sharing among relatives by referring to social relations (a social factor). She also tried to explain food sharing among non-relatives by referring to a cultural value (i.e., a world view).

Fienup-Riordan (1983) observed the Yup'ik on Nelson Island in southwestern Alaska, emphasizing that the subsistence activities of the Yup'ik and related ideologies remain central to their lives despite drastic social changes in their communities. She discovered that the Yup'ik rely on each other through sharing and redistribution within the family network. The Yup'ik give others hunted food as a special ritual gift or share it as part of everyday meals. Fienup-Riordan reported that the Yup'ik can continue sharing only by engaging in hunting and fishing and are able to reproduce social relations between families only through hunting and sharing. Moreover, while it is true that the Yup'ik can obtain food through sharing, they consider it important to give food to relatives and the community so that they can fulfill their obligations (Fienup-Riordan 1983: 357).

1.8.2 Food Sharing in Arctic Canada

Van de Velde and Balikci explored the sharing of seal meat among Netsilik Inuit hunting partners (Van de Velde 1956; Balikci 1970). Damas compared food sharing between three Inuit groups in the central Arctic (Damas 1972a). There have also been studies on food sharing in the central and eastern Arctic (Kishigami 1995, 2000, 2004a). Wenzel investigated food sharing in the eastern Arctic (Wenzel 1995, 2000). Another study was done on food sharing in the western Arctic (Condon, Collings, and Wenzel 1995). Additionally, an investigation summarized the descriptions and perspectives of anthropologists who specialized in food sharing among the Inuit (Lévesque et al. 2000). Another work created a record of the Inuit's discourse in relation to food sharing (Bennett and Rowley 2004: 86–94).

Father Van de Velde engaged in missionary work in the Netsilik Inuit community around Pelly Bay (currently Kugaaruk) in central Arctic Canada and reported on the sharing of seal meat among seal hunting partners in the winter (Van de Velde 1956). Netsilik Inuit hunters used the names of a seal's body parts when referring to their hunting partners, such as "the shoulders" (*aksatkolik*) and "the neck" (*kinguserk*). Each hunter had up to 14 or 15 partners. When a hunter caught a seal and brought it back to his igloo (snow house), his wife or mother slaughtered it and shared the body parts with his hunting partners. In turn, the hunter received the parts of seals hunted by his partners. In the Netsilik Inuit community, two specific hunters would exchange specific parts of a seal. Hunters chose their partner(s) from distant relatives and those outside their extended family. The Netsilik Inuit form a camp group with a large number of non-relatives in winter. Hunters' main subsistence activity during winter was the group hunting of seals at breathing holes. The Netsilik Inuit exchanged parts of seals among partners from outside the extended family. In this sense, they formed social ties across different extended families, thereby successfully distributing food throughout the entire winter camp (Balikci 1970: 135–138). However, the Netsilik Inuit no longer use this system because fewer hunters are engaged in the hunting of seals due to an increased use of rifles and freezers (Buijs 1993). I did not observe the system of sharing seal meat among specific partners at Pelly Bay in the 1990s. In this village, food was shared most frequently among members of extended families (Kishigami 1995).

Damas (1972a) compared food sharing between three Inuit groups in the Arctic region of Canada: the Copper Inuit, the Netsilik Inuit, and the Igloolik Inuit. Damas found that the Copper Inuit used the following three methods in sharing seal meat: *piqatigiit* (sharing between exchange partners), *payuktuq* (voluntary sharing), and sharing through communal meals. Damas observed the following five types of sharing in the Netsilik Inuit community: *niqaiturvigiit* (sharing between exchange partners), *payuktuq* (distributing meat to other tents and dwellings), *minnak* (giving meat to visitors), *akpaaqtauyuk* (inviting selected guests to a meal) and *ningiq* (the sharing of meat by hunters). Damas witnessed the following five kinds of sharing among the Igloolik Inuit: *ningiq* (the sharing of large animals), *payuktalik* (the distribution of meat to other tents and dwellings), *umiaqqatigiit* (sharing among crews of large boats), *nirriyaktuqtuq* (communal feast with the entire village community) and *akpaallugit* (inviting selected guests to a meal). Damas revealed distinctions among the three groups, some of which were due to geographic differences. He noted there were differences in the seal meat sharing partnership system and the sharing among crews of large boats among the three groups (Damas 1972a).

Moreover, Damas (1972a) examined the three types of food sharing that were common among the three groups. His results provided several fascinating findings. First, not only kinship, but also locality was at play in food sharing, since kinship may or may not be related to food sharing (Damas 1972a: 222, 235). Second, the three types of reciprocity (generalized, balanced, and negative) were seen in the three Inuit groups (Damas 1972a: 237). As pointed out by Sahlins, it is probable that there was a correlation between social (family) distance and reciprocity (Damas 1972a: 238). Additionally, in the Igloolik Inuit community, redistribution by the *issumataq* (the head of a hunting camp group) is practiced (Damas 1972a: 237). Third, food sharing among the Inuit had a function similar to insurance (i.e., ensuring survival during food shortages), yet also helped even out the amount of food owned by individuals when food was available in abundance (Damas 1972a: 236).

Wenzel, who conducted research in Kangiqtugaapik (Clyde River) in the northeast of Baffin Island, Canada over the course of 40 years, reported on food sharing and changes in the local community (Wenzel 1991 1995, 2000). Wenzel found that people in the village practiced six types of food sharing: (1) sharing among hunters who do not have family ties and who have worked together in hunting and fishing; (2) sharing between a hunter and the leader of the extended family to which the hunter belongs; (3) sharing between the leader and members of the extended family; (4) sharing between the leader of the extended family and the entire community; (5) sharing between a hunter and the leader of another hunting group that does not have family ties with the hunter; and (6) sharing between leaders of local churches or local cooperatives and the entire community. Wenzel insisted that types (2), (3), (4), and (5) are traditional sharing, whereas types (1) and (6) are modern forms of sharing practiced since the 1950s, when the Inuit settled in Clyde River (Wenzel 1995). In this Inuit community, members practiced *ningiqtuq*—a type of food sharing—prior to settling down in the current village. *Ningiqtuq* is a method of maximizing access to resources by individuals and households; it was practiced in accordance with the order of genealogy of the extended family as well as the age-based order of respect and seniority (Wenzel 2000: 63).

Further, Wenzel reported that the scope of *ningiqtuq* expanded beyond each extended family and applied to the entire Inuit community in Clyde River (population 540). However, Wenzel underscored the continuation of (and changes in) their food sharing system and stated that the *ningiqtuq* system is partially collapsing while these Inuit people still practice a system in which *issumataq* (the head of an extended family group) supervises food sharing within the family (Wenzel 1991: 132). Such shifts took place because the European Economic Community (EEC) introduced a new regulation in 1983 that banned the import of seal furs to EEC countries. Due to the demise of the fur market, the Inuit were forced to exit the seal fur business and lost one of their sources of cash income (Wenzel 1991, 1996, 1995). This led them to experience difficulties in purchasing gasoline and other necessary resources for hunting; therefore, they lost the capacity to continue to engage in hunting and fishing. As a result, Inuit hunters could no longer bring back extra meat that could be shared with other members of the village community (Wenzel 1991: 132). Thus, the quantity and frequency of food sharing in the village decreased. Instead, according to Wenzel, people started participating in communal feasts held by a local church with the entire village community and exchanging presents on Christmas and birthdays (Wenzel 1991: 132).

The ban of the import of seal furs by the EEC in 1983 gave rise to a new sharing system in Clyde River. The gap between those with cash incomes and those without widened; the latter increasingly practiced demand sharing, exerting pressure on those with incomes to share their money (Wenzel 1995, 2000).

Condon researched the Copper Inuit in Holman, in Canada's Northwest Territories, noting that fewer households collaborated with each other and the amount of food shared within the Inuit community declined due to changes in collaboration and modes of sharing. Condon explained this was because individualism became part of people's wage labor and subsistence activities as economic stability increased (Condon 1991: 273). Although the Inuit in Holman now practice food sharing less frequently, they do still share food among households. The practice still plays an important role in the village's socio-economic integration because households with elderly people or widows still obtain meat and fish through food sharing. Condon and colleagues stated that in Holman, members of the community engage in food sharing practices based on generalized reciprocity and balanced reciprocity. Households that actively engage in hunting and fishing share food on the basis of balanced reciprocity with other households in similar situations. They do so on the basis of generalized reciprocity with households that do not actively engage in hunting and fishing. According to Condon and colleagues, the people in the community usually engage in generalized reciprocity when they share food with relatives and friends, whereas they normally engage in balanced reciprocity when they share food with socially distant others. These two kinds of food sharing contribute to the integration of the community (Condon, Collings, and Wenzel 1995: 41–43). They found that food sharing in the community was determined by the following factors: (1) the amount of meat and fish caught and shared within the community; (2) the quantity of meat and fish that the giver has received from other households; (3) the number of relatives on the giving side; (4) the likelihood that people on the giving side will catch animals; and (5) the degree of the hunter's generosity (Condon, Collings, and Wenzel 1995: 41).

Searles conducted research in Iqaluit, Baffin Island, and found that Inuit individuals who regularly contribute food and resources for communal purposes have significant power and authority (Searles 2002: 61–62). For the Inuit, food sharing is a tool to obtain respect and prestige. On the other hand, Searles also reported that a person who fails to share food will be the subject of denigration and belittlement (Searles 2002: 62–63).

Graburn, who conducted research in Salluit, a village facing the Hudson Strait, in the late 1950s, found that rules for sharing catch in the town varied depending on the size of the catch and the amount of the meat (Graburn 1969: 66–67). The rules that Graburn discovered were as follows: (1) large animals with a weight exceeding 250 pounds (approximately 115 kg) such as beluga whales, walruses, bearded seals, and harp seals should be shared among all members of the camp or relatives by any available ways. (2) Animals weighing between 50 pounds (approximately 23 kg) and 250 pounds (approximately 115 kg) such as caribou, ringed seals, young bearded seals, and young harp seals should normally be shared among close relatives within the extended family. (3) Small animals (such as birds and rabbits) and eggs should be kept by the household of the hunter or gatherer. (4) If the amount of small animals hunted and eggs collected is equivalent to that given in rules (1) and (2), the small animals and eggs should be shared in accordance with rules (1) and (2). Additionally, Graburn pointed out that the Inuit shared food even when it was not necessary to do so because food sharing helped them maintain ties with family and friends (Graburn 1971: 108). Graburn also affirmed that sharing and visiting were key to maintaining collaborative relationships among different family groups as well as alliances with neighboring groups (Graburn 1971: 110).

Riches carried out fieldwork in Port Burwell (Killiniq), a community located at the north end of the border between Quebec and Labrador, from the 1960s to the 1970s. In Port Burwell, members of the community were obliged to give not only to their relatives but also to neighbors who were not relatives. Food was shared first with relatives, followed by neighbors. Food was not shared with any individuals other than close relatives during food shortages (Riches 1981: 220, 222, 224). Riches attempted to elucidate why the Inuit consider themselves obliged to give to (share with) other Inuit people. To explain the reproduction of obligations, he used the concept of the multiplex relationship coined by Gluckman (1955). In small societies such as that of the Inuit, a small number of people form a camp in which relatives and neighbors closely interact and share multiple interests in their everyday life/lives. Social relations constructed within a camp are related to various important interests. For members of the camp to survive, they must maintain social relations. From a long-term perspective, it is more effective in terms of adaptation to uphold social relations (by giving out food) than ruining them (by not doing so). For these reasons, the idea of the "obligation to give" food is reproduced in this Inuit community (Riches 1981).

Dorais conducted fieldwork in Quaqtaq, a village located at Ungava Bay, and found that *ningirsiniq* (sharing) and *ikajuutiniq* (mutual aid) are basic characteristics of Inuit culture (Dorais 1997: 56). He also revealed a gap between reality and the Inuit's discourse surrounding food sharing. Put differently, while the Inuit believe they should help (and share with) any individuals, they are in fact only helping each other within each extended family or specific kindred (Dorais 1997: 68).

In the Nunavik region in Arctic Quebec, Canada, the Inuit frequently engage in unidirectional food sharing within the extended family. The James Bay and Northern Quebec Agreement in 1975 led to the launch of a hunter support program. Consequently, people have been gradually adopting a new practice of food sharing (Kishigami 2000, 2004a). Gombay examined the impact of commodifying country food on the contemporary Nunavik Inuit community in Puvirnituq, Nunavik, from the 1990s to the early 2000s (Gombay 2004, 2005). Chabot carried out a comparative study of the household economy among three villages in Nunavik (Chabot 2001, 2003), and reported that food sharing is still frequently practiced in modern Inuit society despite undergoing drastic changes.

1.8.3 Food Sharing in Greenland

Research on Inuit communities in Greenland includes studies on food sharing in Ammassalik and southeast Greenland (Robbe 1994: 334–359; Hovelsrud-Broda 2000); those on food sharing in Inuit communities in northwest Greenland (Nuttall 1991; Dahl 2000: 175–185); and those on changes in Greenland's distribution/sharing system (Petersen 1989; Marquardt and Caulfield 1996).

Hovelsrud-Broda conducted fieldwork in Isertoq, a community in southeast Greenland. In Isertoq's Inuit community, a household is the basic unit; people engage in two types of food sharing: within the household and between households. Food sharing among the households can be further divided into five kinds. The first entails one household sharing (country) food with another. The second type involves a party held to celebrate children's birthdays and their first participation in hunting. The third kind refers to a bidirectional exchange through which food is regularly shared between certain households. The fourth type is the exchange of polar bear meat between hunting partners. The fifth type is the sharing of a minke whale with the entire village community (Hovelsrud-Broda 2000: 204–206). While people engage in the first and third forms of sharing daily, they are practiced within a network of close ties such as blood relatives, in-laws, and pseudo-relatives (Hovelsrud-Broda 2000: 199). Hovelsrud-Broda pointed out that the transfer of country food (such as seal meat) helps to maintain and enhance family relationships in the Inuit community. Moreover, the Inuit in Isertoq highly value country food (especially seal meat) as a cultural food resource, and prove themselves as members of the Isertoq Inuit community by eating such food (Hovelsrud-Broda 2000: 211). Only country food is shared across different households in the entire village. Sharing helps food resources to be distributed throughout the community (Hovelsrud-Broda 1999: 47).

Nuttall carried out fieldwork on food sharing in Kangersuatsiaq, a settlement in northwest Greenland (Nuttall 1991, 1992: 136–154). In Kangersuatsiaq, members of the community are obliged to give meat (especially seal meat) to others; no hunters refuse requests for meat (Nuttall 1991: 219). Nuttall analyzed food sharing in Kangersuatsiaq by drawing on the relationship between food sharing and people's world views, social relations, and the community. These Inuit people believe that animals (such as seals and beluga whales) give their lives to hunters, and hunters who have hunted an animal should therefore not monopolize it. They believe that hunters are obliged to share hunted animals with others. Moreover, Nuttall explained that in Kangersuatsiaq, absolute rights over animals hunted by a member of the community are held by the community. Hence, hunters are compelled to give out animals they have hunted. In many cases, hunters give others their catch without expecting reciprocity. In turn, hunters acquire prestige and a reputation that they are admirable hunters (Nuttall 1991: 219–220). Further, in this community, people approve, enhance, and demonstrate existing and future social relations by sharing meat. They also approve and reproduce perspectives of the world (values) and the community (Nuttall 1991: 221). Nuttall drew on the case of Kangersuatsiaq in criticizing the alliance hypothesis (Damas 1972b), which explains that the sharing of meat helps maintain relationships among different groups of relatives. Meanwhile, Nuttall underscored that unlike beluga whales and other animals, seal meat is not traded with cash, asserting that while the sharing and distribution of seal meat can contribute to maintaining community ties, selling and buying only benefits the seller and buyer (Nuttall 1991: 220).

Dahl, who conducted research in Saqqaq, a settlement in northwest Greenland, insisted that the giving of meat as a gift among the Inuit should be distinguished from sharing (Dahl 2000). In Saqqaq Village, people practice three types of sharing: *ningerpoq* (the sharing of traditional catch), *agguarpoq* (the sharing of beluga whales for which regulations have recently been introduced among boat crews), and *pajuppoq* (giving meat as a gift). The first two types are part of production relations, whereas the third one is part of social relations or the community's social exchange system. In the cases of the first two kinds, for example, hunters who have participated in hunting activities share a beluga whale in accordance with rules derived from their rights for specific parts of the whale, certain amounts of meat, and *mattak* (skin part with fat). For the third type, hunters voluntarily share meat on the basis of social norms without expecting reciprocity. In other words, the practice of sharing in Saqqaq can be broadly divided into two types: sharing in accordance with rules based on production relations, and voluntary sharing that does not involve obligations (Dahl 2000: 175–178). Moreover, Dahl found that beluga whale hunting in Saqqaq Village involves the entire village community, explaining that boats taking part in beluga whale hunting will equally receive meat irrespective of the boat's size in Saqqaq. In contrast, the portion that hunters receive is determined on the basis of the boat's size in Illulissat Village (Dahl 1989: 33). These findings suggest a growing difference in *agguarpoq* between different areas as regulations for *agguarpoq* have recently been introduced.

In Greenland, meat and *mattak* of beluga whales are traded with cash. This practice has led to changes in the system of sharing and distribution. In the past, households were the socio-economic unit of Greenland's Inuit communities. When a household lacked food, it could receive food from other households in the community. However, if a household received objects other than food, it needed to give back another object or service as a reciprocal gift. Meanwhile, people from different communities exchanged items (Petersen 1989: 81). Food sharing within the community not only generated a sense of unity and friendship, but also produced community well-being (Petersen 1989: 83). Traditional Inuit communities in Greenland, which comprised hunter households, have changed from homogeneous to heterogeneous communities. Put differently, in most modern communities, members are divided into two groups: three-quarters of the community engage in wage labor, whereas one quarter takes part in hunting and fishing. The mainstream of food sharing in the community used to be based on generalized reciprocity. However, the Inuit currently find it necessary to involve cash in the sharing channel, as their community is becoming increasingly heterogeneous. Petersen claimed that these two heterogeneous groups (wage laborers and subsistence hunters/fishermen) can coexist when wage laborers purchase country food (the meat of locally hunted land and marine animals and fish) from hunters and fishermen (Petersen 1989: 87).

Denmark colonized Greenland in 1721. Since then, the Inuit in Greenland have been selling country food to Danish settlers. Therefore, the buying and selling of country food has a long history. Notwithstanding, the Inuit have rarely traded country food within their own communities. Since 1988, Greenland's self-government has encouraged the exchange and sale of all types of country food meat to reduce Greenland's dependence on food imports. Country food is sold in Inuit communities as follows: (1) through individual transactions to households; (2) through individual transactions to institutions such as schools, hospitals, and aged care facilities; (3) to small businesses or at a public market; and (4) to government-owned processing factories. Marquardt and Caulfield found that hunters in Greenland sell country food not to earn cash, but to maintain their lifestyle and subsistence activities (Marquardt and Caulfield 1996).

1.8.4 Similarities and Differences in Food Sharing in the Arctic Regions

Thus far, I have provided an overview of food sharing among the Yup'ik (Yupiit), Iñupiat, and Inuit who live in the Arctic regions from Alaska to Canada and Greenland. In the studies explained in this section, researchers have adopted different perspectives and methods of description. A comparison of these studies can identify similarities and differences among areas and groups.

I will show and explain the similarities between the Yup'ik, Iñupiat, and Inuit. First, while the cash economy has penetrated the modern Arctic regions, food sharing is still practiced regularly.

Second, modes of food sharing include giving, exchange, and redistribution. Third, most food is shared among relatives. Fourth, while large animals and many small animals are shared, a small animal or small amount of catch is not shared. Fifth, food sharing is closely interrelated with people's world views pertaining to the relationships between humans and animals. Sixth, studies have found the following characteristics of food sharing: providing insurance, maintaining social ties, helping to obtain prestige and reputation, and contributing to the well-being of the community or/and extended families.

A difference between areas is that while country food is widely traded with cash in Greenland for historical reasons, it is rarely traded with cash in other places such as the Arctic regions of Alaska and Canada. Additionally, in Canada and Greenland, state policies and agreements with indigenous people have given rise to new forms of food sharing and exchange. Differences in ways of incorporating the market economy and state policies may have affected food sharing practices in the Arctic regions.

1.9 The Potential Contribution of Studies of the Inuit, Yup'ik, and Iñupiat to Food Sharing Research

This section explores the potential contribution of studies on hunter-gatherers in the Arctic regions to those on food sharing among hunter-gatherers. It is evident that case studies of food sharing among the Inuit, Yup'ik, and Iñupiat help examine existing hypotheses on food sharing. Elsewhere in this chapter, I have pointed out that social network analysis and gender studies have enabled investigations on the Inuit to contribute to food sharing research. Inuit studies can also contribute to other aspects. Bird-David (2019) stressed that studies on food sharing in hunter-gatherer communities have issues with scale and lack a perspective on kinship. I believe that Inuit studies can help to address these two issues and also shed light on diachronic changes in food sharing.

1.9.1 Differences in Food Sharing Between Small and Large Groups

Imamura (1993) and Kitanishi (1997), who studied African hunter-gatherers San and Aka respectively, examined food sharing through negotiations based on face-to-face interactions. They reported that food sharing behavior varied depending on the camp's population. In a camp with a small population, members of the camp shared food not based on kinship, but on close face-to-face interactions which they shared in every aspect of life, including hunting activities. Meanwhile, in a large settlement or camp with a large population located near a settlement, members of the settlement or camp were selective in determining with whom they shared food. Unlike small camps, people did not share food with every member of the settlement or camp. Kinship and physical distance were key factors influencing the decision of recipients in food sharing practices in a place with a large population. Given such cases, there are differences between small groups where people share every aspect of life and large groups formed with multiple small groups in terms of the target, frequency, and characteristics of food sharing.

When the population of the Australian Aboriginal Pintupi increased due to moving to a settlement, their sharing network ceased to function. Myers explained that the Pintupi stopped sharing food in the community because an excessive number of people demanded sharing and there was no longer pressure to practice reciprocity (Myers 1988: 58–59). As detailed elsewhere, Winterhalder found that the effect of food sharing is maximized in a small group with 7–8 hunters (Winterhalder 1986a, 1986b). Smith asserted that in a large group in which few people engage in face-to-face interactions, many people store food; hence, the effect of food sharing will be lost (Smith 1988).

In what follows, I show two hypotheses that outline why all members of a community frequently practice food sharing in a group with a small population, where face-to-face interactions form the foundation of the community. Ingold argued that sharing is practiced on the basis of "the relationship between individuals who engage in direct face-to-face interaction in a closed social group" (Ingold 1988: 283). I consider that Ingold's opinion suggests that everyone in a small camp share food while sharing life experiences and engaging in collaborative work irrespective of kinship. Tanno (1991), Imamura (1993; 1996), and Kitanishi (1997) also supported this idea. Thus, sharing every aspect of life leads to frequent food sharing involving all members of a community.

The other hypothesis explains why hunter-gatherers practice food sharing in an obligatory manner. As Riches (1981) put forward the hypothesis, members of a small camp work together closely and cohabitate while sharing multiple interests. Social relations are related to the interests of the group's members; maintaining social relations is beneficial to their survival. When considering long-term effects on adaptation, it is more effective for people to maintain social ties by temporarily providing food than to break them by not doing so; therefore, the idea that giving is an obligation is reproduced in hunter-gatherer societies.

It is important for researchers to elucidate why people share food in various manners in different communities with large populations where multiple camps have settled, whereas all members of an independent camp with a small population outside a village share food with each other.

Case studies of Inuit communities from the 1980s to the 2010s found that families that jointly engaged in camping and hunting shared food in almost any situation, irrespective of kinship. However, in a large village, meat brought back to the village by hunters is usually shared and distributed in accordance with kinship. Significant differences in sharing among the Inuit have been observed between campsites and villages. Hence, examining such differences may contribute to research that sheds light on the relationship between food sharing and group size.

1.9.2 Food Sharing and Family/Kinship

Anthropological studies of hunter-gatherer societies in Arctic North America differ from those in other areas in that the former have focused on family and kinship. Studies on the Inuit, Yup'ik, and Iñupiat have verified that family and kinship play important roles in subsistence activity and social structure. These findings also apply to practices of food sharing (e.g., Bodenhorn 2000; Burch 1975, 1988; Damas 1963, 1972; Kishigami 1995, 2000, 2004; Ready 2016; Wenzel 1981, 1991, 2000; 2019).

Thus, conducting more research on food sharing in the Arctic region could help to explain the relationship between food sharing and family/kinship. Particularly notable is that Sillander (2019) pointed out the significance of social closeness in food sharing practices in small scale groups. Based on findings to date, studies on the Inuit, Yup'ik, and Iñupiat can help to answer this question.

1.9.3 Diachronic Changes in Food Sharing

It is true that there have been changes in food sharing among the Canadian Inuit due to their articulation with the market economy and sedentarization since the 1960s. On the other hand, the Inuit still frequently engage in food sharing in the modern context of globalization.

A summary of existing studies indicates that hunter-gatherers have continued to practice food sharing in the modern socio-economic environment even as it has changed with time. A few studies have reported that food sharing among these people has socio-economic effects. However, there is a significant dearth of research on the following elements: reasons for the diachronic practice of food sharing in certain study sites; how this practice has changed; how food sharing is practiced in the modern socio-economic setting; and what kinds of functions and effects food sharing has in the modern socio-economic context.

Several studies have examined food sharing and its historical transformation in the Arctic regions, including studies of the Yupiit on the Chukotka Peninsula and Alaska, the Iñupiat in Alaska, and the Inuit in Canada and Greenland. These investigations have accumulated findings related to food sharing among specific groups in given areas (e.g., Van de Velde 1956; Damas 1972a; Langdon and Worl 1981; Fienup-Riordan 1983; Burch 1988; Nuttall 1991; Bodenhorn 2000; Kishigami 1995, 2000; Wenzel 1981, 1995, 2000; Hovelsrud-Broda 2000; Levesque et al. 2000). I believe that studies on long-term changes in food sharing practices in a selected indigenous village in the Arctic region will help to shed light on diachronic changes in food sharing among hunter-gatherers.

1.9.4 The Objectives of This Book

As discussed thus far, I believe that testing existing hypotheses on food sharing among hunter-gatherers in the Arctic regions (such as the Inuit), as well as exploring the research questions below, can contribute to elucidating food sharing among hunter-gatherers.

(1) How do camps with a small population and villages with a large population differ in terms of food sharing practices?
(2) What is the relationship between food sharing and family/kinship?
(3) How has food sharing diachronically changed in a selected group of hunter-gatherers? What is the current situation of food sharing in that group?

I carried out research concerning kinship and food sharing among the Inuit in Akulivik Village, Nunavik in Quebec, Canada from 1984 to 2016. I also observed whale hunting by the Iñupiat and their sharing of whale meat in Barrow Village (current Utqiaġvik) in Alaska from 2006 to 2013. This book aims to verify several hypotheses related to food sharing and to answer the above-mentioned research questions. In doing so, the book draws on findings from my fieldwork to date.

Chapter 2
Food Sharing in the Inuit Society: A Case Study from Akulivik in Nunavik, Canada

2.1 Introduction

The Inuit, Iñupiat (Iñupiaq), and Yupiit (Yup'ik) peoples have been the principal occupants of much of the arctic tundra region of North America lying beyond the continental tree line. They lived in self-sufficiency by making full use of the natural resources available in their territories and trading some of these resources with neighboring groups before they came into close contact with Europeans and others starting roughly in the sixteenth century.

The Inuit are indigenous hunter-gatherers in the Arctic of Canada. It is known that food sharing is one of the prominent characteristics of their social practices. In this chapter, I will focus on the food sharing practices in the Inuit community of Akulivik in Nunavik (arctic Quebec), Canada. Then, I will delineate several features of food sharing among the Inuit and discuss the social, economic, and cultural significance of these practices for them.

I will outline several cases of Inuit food sharing practices primarily in the 1990s in Akulivik, where I conducted a series of my field studies from 1984 to 2016.

2.2 History of the Nunavik Region and Outline of the Akulivik Society and Economy

The village of Akulivik lies on the eastern coast of Hudson Bay (60°, 48 min, north latitude and 78°, 8 min, west latitude, see Map 2.1). It was officially established as a municipality of the Kativik regional government in Nunavik, Province of Quebec, Canada in 1976. I shall now briefly outline the modern history of the Nunavik region, Canada, and describe the social and economic aspects of the village in the 1990s.

N. Kishigami, *Food Sharing in Human Societies*, Trust 4,
https://doi.org/10.1007/978-981-16-7810-3_2

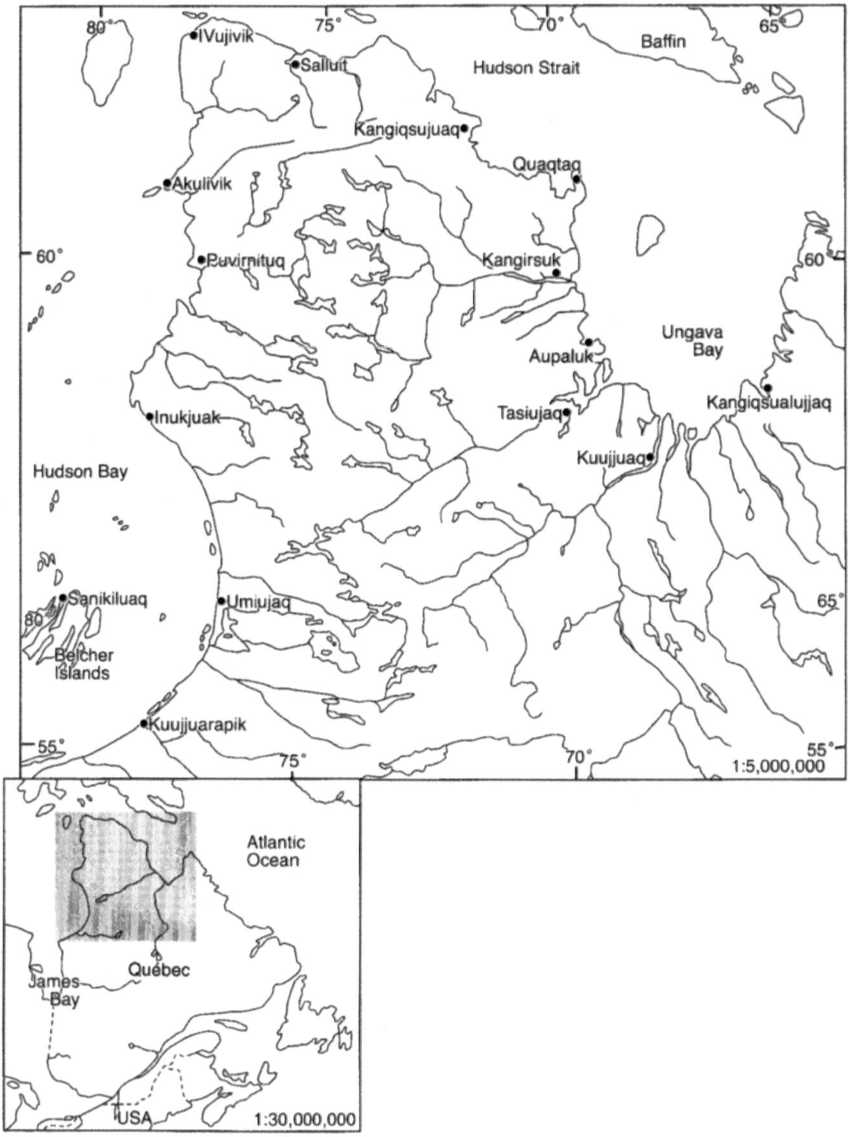

Map 2.1 Nunavik, Canada

2.2.1 Recent Inuit History of the Nunavik Region in Canada

Before the middle of the twentieth century, several factors combined to bring about substantial changes among the Inuit in Canada because of close contact and interaction with Europeans, Euro-Canadians, and Americans. These factors include involvement in the fur trade, Christianization, and a drastic population decline due to the contraction and spread of infectious diseases because of increasingly intensive contact with non-Inuit people.

In Nunavik, the Inuit began trading furs of arctic fox and other animals with Europeans and Euro-Canadians around 1910. By the 1930s, most Inuit had been converted to Christianity by either Anglican or Catholic missionaries (Kishigami 1994: 228–229). Further, from the late 1930s to the end of the 1940s, tuberculosis prevailed at almost epidemic levels among the Inuit in this region. Thus, the government of Canada was forced to intervene medically.

A turning point in Nunavik Inuit life occurred around 1950. Until the middle of the decade, the Inuit in Nunavik were non-sedentary and rarely occupied a particular place for a prolonged period. However, around the early 1950s, there began the uneven struggle between the attraction of government settlements, which had better houses and easier access to medical, church, trading, and welfare services, and that of the land, which offered traditional pursuits, but also posed the uncertainty of disease and starvation (Paine 1977b: 13). Additionally, the government wanted to assimilate the Inuit into the Canadian mainstream both culturally and socially. Therefore, in the interests of the efficiency of administrative and social services, the federal government of Canada planned and began to implement the sedentarization policy of the Inuit population in the Arctic region around the mid-1950s (Damas 2002). The Inuit living in the Cape Smith (Akulivik) area were thus moved during the 1950s and 1960s to Puvirnituq (a regional center settlement on the eastern Hudson Bay coast of Arctic Quebec). By the mid-1960s, Inuit people drawn from different Cape Smith camp groups lived in Puvirnituq through the year and became increasingly involved in wage employment and the transfer economy. As the Inuit became more dependent on the national economy of Canada, money became ever more significant in their daily life. The main income sources at this time were transfer payments by the federal government, petty wages from labor, and cash from selling soap stone carvings and furs of seals and arctic foxes.

From the late 1960s to the early 1970s, government protectionist and assimilation policies targeting Indigenous peoples of Canada, including the Inuit, a process which Robert Paine (1977a: 3) has termed "welfare colonialism," were implemented in the Arctic region by the federal authorities. However, after the land claim movement gained momentum in the early 1970s, the federal government undertook the negotiation of several land claims with representatives from each of the concerned Indigenous groups in the region.

It was in 1971 that the Premier of Quebec, Robert Bourassa, declared the "James Bay Hydro-Electric Project" to be an important development not only for Nunavik and the Cree regions of northern Quebec, but also for the province as a whole. As the

Inuit and Cree voiced their opposition to the project, the Quebec and federal governments began to negotiate Indigenous rights, including land rights. In November 1975, all parties agreed to and signed the "James Bay and Northern Quebec Agreement" (JBNQA).[1] While both the Inuit and the Cree surrendered their aboriginal title and rights as a result of the agreement, they did obtain specific rights to subsistence resources and limited lands, as well as compensation money. Further, JBNQA also provided the means for a hunter support program, which was initiated in Nunavik in 1983.

Canadian Inuit seal hunting was compatible with the cash economy until the early 1980s (Wenzel 1991), with seal hunting task groups remaining organized based on kinships and with meat being shared in particular ways within an extended family or with Inuit of different families. Thus, through the customary sharing and giving practices, social relationships between givers and receivers of seal meat (major country food) were maintained, while sealskins were sold by the Inuit to the Hudson's Bay Company or Co-Op (Co-Operative) for cash. This cash was then used to purchase ammunition and gasoline for snowmobiles and outboard engines, which were necessary materials for carrying out contemporary hunting and fishing activities. In a real sense, therefore, Inuit hunting and fishing activities that received cash investments and their sharing practices enabled the reproduction of socioeconomic relationships in the contemporary context. This shows that Inuit subsistence activity is a socioeconomic system, composed of procuring and sharing and organized by special social relationships, such as kinships and hunting partnerships (Wenzel 1991, 2013, 2019). Furthermore, these activities are correlated with indigenous knowledge, worldviews, and ethno-technology.

The symbiotic relationships between subsistence activities and the fur trading economy continued until 1983, when the European Community (EC) banned imports of specific animal products, especially seal products. This event resulted in the collapse of the fur market in Europe and, in turn, led to a drastic decrease in income from Inuit hunting activities as hunters found it harder to meet their costs. The resulting decline in hunting and fishing activities has since caused great changes in social relationships upon which Inuit subsistence is based.

2.2.2 *The Akulivik Economy and Society*

After a 15-year stay in Puvirnituq, the Inuit of the grandparents and parents' generations of the Akulivik people returned to their homeland, the Cape Smith region, in the early 1970s and established the village as a municipality in 1976. This part describes the economic and social structures of the Inuit community in Akulivik in the 1990s.

2.2.2.1 Economic Structure

The contemporary economy of the Canadian Inuit is characterized by a mixed economy of subsistence and cash (Wenzel 2019). The Nunavik Inuit have been politically absorbed into the Province of Quebec and Canadian governmental systems and economically integrated into both the world system and the national economy of Canada since the 1950s. As a result of their experience over the last five decades of resettlement, the Nunavik Inuit could not fully carry out hunting and fishing activities without an array of imported technologies. Thus, cash was used to purchase gasoline, ammunition, rifles, nets, snowmobiles, and boats with outboard engines.

In the contemporary Inuit village of Akulivik in Nunavik (see Map 2.1), the following local wild animals, among others, were used as food resources: ringed seals (*natsiq*), bearded seals (*ujjuq*), beluga whales (*qilalugaq*), walrus (*aiviq*), polar bear (*nanuq*), caribou (*tuttuq*), arctic char (*iqaluppik*), white fish (*kavisilik*), lake trout (*isiuralittaaq*), ptarmigan (*aqiggiq*), Canada geese (*nirliq*), snow geese (*kanguq*), and eider duck (*mitiq*). The Inuit also harvested birds' eggs, berries, seaweed, shellfish, and sea urchins. The annual harvesting cycle of the Akulivik Inuit is summarized in Fig. 2.1.

Although almost all adult Akulivik men engaged in hunting and fishing in the 1990s, only about half of the 60 adult males in the contemporary community were active hunters and fishermen who did not participate in wage employment as their main source of income. Participation in wage labor, however, was generally low because few jobs existed in Akulivik or other Nunavik Inuit villages. For example, in 1995, there were 55 full-time jobs, 34 part-time jobs, and 24 mainly seasonal

Month(1-12=January to December)

	1	2	3	4	5	6	7	8	9	10	11	12
game												
Arctic Char	○	○	○	○	○	○	◎	◎	○	○	○	○
Land-locked Char	◎	◎	○	○	○	○	○	○	○	○	◎	◎
White Fish	◎	◎	○	○	○	○				○	◎	◎
Riged Seal	○	○	◎	◎	◎	○	○	○	○	◎	○	○
Bearded Seal	○	○	○	○	○	○	○	○	○	◎	○	○
Beluga Whale						○	○	○	◎	◎		
Walrus								○	◎	○		
Polar Bear	◎	◎	○	○	○	○	○				◎	◎
Calibou	◎	◎	◎	◎	◎	◎	○	○	○	○	○	○
Snow Geese						○	○	○	○			
Canada Geese					○	○	○	○	○			
Eider Duck				○	○	○	○	○	○	○		
Birds' Eggs						○						
Ptarmigan	○	○	○	◎	◎	○	○	○	○	○	○	○
Berries								○	○			

Game means 'primary' game. ○ means a harvesting month. ◎ means the best harvesting month.

Fig. 2.1 Annual cycle of hunting and fishing in Akulivik

construction jobs in Akulivik, whose population was approximately 400 that year (Lefebvre 1996:158, 184). More women tended to keep fulltime jobs for a longer period than men. The population obtained its food from hunting/fishing (far less than 50%), with the remainder coming from imported southern foodstuffs purchased from the Co-Op and other stores.[2] The men who were active hunters, however, captured most of the seals, caribou, and fish that entered the village depending on transfer payments from the federal and provincial governments and/or sold soapstone carvings to obtain funds to support their hunting and fishing activities. Thus, this shows "men as the hunters, and women as the wage workers."

2.2.2.2 Social Structure

Before the 1950s, the general pattern of Cape Smith (Akulivik) Inuit subsistence activities was organized around small camp groups, which seasonally migrated from one place to another in the northwestern Nunavik region. Summer and winter camp groups assumed two different social configurations. Typically, summer camp groups were made up of one large extended family, the nuclear domestic units of which resided in several separate tents. The winter group, on the other hand, was made up of several summer groupings. Akulivik Inuit summer groups usually ranged in size from five persons (1 tent) to approximately 20 persons (4 tents) in the mid-1930s before the relocation to the Puvirnituq area. While summer camps were small, the winter camp groups were large with a maximum of 140 persons and established on the sea ice. Hunting and other seasonal activities were conducted by hunters of the camp groups and followed seasonal changes in important animal and fish resources.

By the late 1960s, virtually all Nunavik Inuit were established in sedentary government villages. These communities were usually composed of people from formerly different camp groups (extended families), which, prior to government village life, had existed independent of economic support from the state. Since the 1960s, however, government villages have become the centers of residential life among the Inuit.

In the past, several distinct extended families co-resided in a winter camp, and the frequency of marriages among these families was high. Due primarily to technological innovations in hunting equipment and tools (e.g., use of high-powered rifles, snowmobiles, canoes with outboard engines) and economic individualization in government village life, hunting groups became smaller, with activities sometimes being conducted by single individuals, although cooperating units made up of father-son pairs, siblings or cousins, and/or friends remained the norm. However, it must be noted that each extended family continued to function as the primary unit for food sharing and provision of mutual help in the village.[3]

In 1996, there were 411 persons living in 87 households in Akulivik, with the average household consisting of 4.7 persons, while the average household size in 1990 was 6.1 persons. Thus, the household size decreased during those 5 years and a nuclear family form of residence best describes the situation of a majority of the village's population.

Most household heads in Akulivik are related to each other by blood or marriage and, in point of fact, genealogically, most household heads share a common ancestor three to four generations removed. Akulivik, in reality, is virtually composed of one large kin group by virtue of the "ancestral" connections that relate most extended families. In terms of daily operations, there were no more than seven large extended families whose elders shared at least one ancestor in four ascending generations in the community in the 1990s.

For the purposes of this chapter, I will distinguish sharing practices that existed in the 1980s from those that developed later because of the Nunavik Hunter Support Program, which was instituted in the mid-1980s (Kishigami 2000). I will call the former "customary food sharing" and the latter "new food sharing" for convenience. Also, there was another form of food sharing through local FM radio announcements. This is also classified as a new practice.

2.3 Customary Food Sharing Practices in Akulivik, PQ, Canada

2.3.1 Outline of Customary Food Sharing Practices in the 1990s

In Akulivik, food items to be shared were the meat of land animals and sea mammals, fish, and many kinds of store-bought food. The process involved (1) an immediate distribution of game among hunters in the hunting group, (2) a distribution of the game between the hunters and other camp members or villagers, and (3) a distribution of the divided meat or other food through meals and/or giving among camp members or villagers.

Hunters usually ate meat or the liver of game at the kill site soon after the hunt or during the butchering. Then, the hunter who killed the animals took the fur and some meat for himself. He then distributed the remaining meat to the other hunters. In most cases, hunters who participated in the hunting or butchering of game would take some meat for their needs, while also taking into consideration the number of hunters and the total amount of meat remaining. Although there had been rules for division of large game such as beluga whale and walrus in the past (cf. Graburn 1969:68–70), the Akulivik Inuit did not abide by such traditional rules in the 1990s.

Once hunters brought meat to their camp or village, they usually personally delivered part or all of it to their close kinsmen (e.g., parents, grandparents, siblings, uncles, aunts, and/or children) in other households. Furthermore, they sometimes gave meat to their neighbors, elders, sick persons, namesake partners (*sauniq*), and ritual midwives (*sanajik*) in the village. Sometimes, Inuit in need of food might visit a successful hunter to obtain some meat. A needy Inuk might also make a telephone call to the local FM radio station to announce his/her need for food when he/she could not get any from kinsmen or neighbors. The Inuit in Akulivik seldom lent food to or borrowed it from others. Indigenous food was regarded by the Inuit as something to be given or received gratis. At camping or hunting sites, the Inuit tended to share food with all those present regardless of whether they had kinship ties with them or not. However, in the village they tended to share food primarily within their extended families.

A hunter who harvested a large animal such as a beluga whale shared *muktuk* (skin parts of beluga) or meat with all the villagers or camp mates even if the pieces were small. In hunts involving large private boats, the boat captain and hunters on board shared the game. Then, if there was surplus, the boat captain personally gave some of it to other villagers who met him on the shore. Meat was also shared with a hunter's kinsmen and neighbors through meals. In Akulivik, many Inuit frequently visited their relatives, especially their parents or grandparents, for lunch and supper or were invited to meals by their kinsmen and others. When a person's parents or grandparents were alive, he or she tended to have his/her lunch and supper at the parents' or grandparents' place. Meals for lunch and/or supper in Akulivik were most often shared with members of the extended family coming from several households.

2.3.2 Customary Food Sharing in Nunavik Society in the 1990s

In the 1990s, the Inuit of Akulivik continued to practice food sharing and mutual help, especially during periods of reduced country food availability. Hunters who participated in cooperative hunting or fishing activities generally shared game on the sea ice or tundra almost immediately after animals were caught. This sharing was called "*ningiktuk*."

When hunters returned to the village, they customarily gave meat directly to their kinsmen and/or neighbors who lived in houses physically adjacent to their own. This sharing was called "*parutuk*." Alternatively, successful hunters invited these people to participate in meals at their houses. This sharing was called "*kaikurik*."

2.3.3 Game Sharing Among Hunters

Since about 1960, due to the use of technological innovations such as snowmobiles and high-powered rifles, Inuit hunters have typically engaged in hunting and fishing activities either individually or in smaller cooperating groups than before. Thus, the frequency and amount of game sharing among hunters, especially in the immediate aftermath of a kill, have decreased since the advent of these introductions. However, Akulivik hunters still considered game sharing to be a principal normative practice. Among cooperating hunters, the individual who captured game directed the transfer of meat to other task group participants. Furthermore, a hunter who captured large

Photo 2.1 Seal meat sharing by Inuit hunters. Akulivik, PQ, Canada October, 1999. By N. Kishigami

game species, such as a bearded seal or beluga whale, gave a portion to his kinsmen and fellow villagers with whom a less formal connection might exist (Photos 2.1 and 2.2).

In order to illustrate the range of sharing at this level, I will highlight four cases involving winter sealing, ptarmigan hunting in winter, and beluga whale and polar bear hunting in summer. Each of these case illustrations are based on data recorded in 1998.

(1) A hunter ("K") left Akulivik alone to hunt seal and fish for arctic char for one day in January 1998. After departing, he encountered four other hunters in the area where he intended to hunt and joined them to form a cooperative seal hunting group. On this occasion, only K was successful, catching a big bearded seal. Following K's success, the meat and viscera from his seal, except for the heart (which was kept by K), were divided equally between K and the other four hunters. In addition, the sealskin was given to one of the unsuccessful men who wanted to have it, and the blubber was apportioned to each hunter according to their stated need. In this way, the hunters of the group shared the game almost equally.

After returning to Akulivik, K made a telephone call to the community FM radio station announcing that he had a bearded seal, inviting anyone who wanted a portion to his house. A number of villagers came to take away parcels of meat. He also

Photo 2.2 A private boat captain giving beluga *maktak* to a villager. Akulivik, PQ, Canada October, 1999. By N. Kishigami

delivered meat to several persons as gifts, and some people also visited K later to participate in meals of the cooked bearded seal.

(2) "A" in Akulivik went alone to hunt ptarmigan and net arctic char at the end of January, catching about 30 ptarmigan and five arctic char. Upon his return to the village, he used the FM radio station to invite the villagers to his house to receive some birds. He also sent his son to deliver some of the catch to the widows and elders who were his neighbors, while his wife made telephone calls to invite specific people to supper. In this way, his catch was distributed to a wide circle of villagers. Persons who came to A's house for a share were A's elder brother (three ptarmigans, one arctic char), a widow who was a school friend of A's wife (one ptarmigan), and one sick elder in bed (several ptarmigans). Persons to whom A's son delivered his catch were the widow of his late cousin (three ptarmigans and one arctic char) and an old couple next door (three ptarmigans). Sharing through meals (three arctic chars and 14 ptarmigans) occurred among household members (A, A's wife, daughter, and son) and others (A's daughter and her son, A's wife's brother, A's hunting partner, A's brother's daughter and her husband and daughter, and A's wife's cousin). The participants in the meal were members of A's extended family.

(3) In June, two parties of three and four hunters, respectively, went after beluga whales and caught several (six in total). The meat and *muktuk* (whale skin parts) were divided and delivered to all the households of the village after each hunter had taken his share.

(4) At the end of July, a young man, "H," killed a polar bear at Cape Smith Island close to Akulivik. It was his first polar bear kill. He butchered the bear under the

instruction of an old man and then divided the meat equally among ten families who had come to the island to fish and picnic. After H's return to Akulivik, he gave the head and skin of the animal to his symbolic midwife ("*sanajik*").[4] When the families who had been at Cape Smith Island returned to Akulivik with the meat they had received, there was more that was distributed to other villagers through meals and as gifts.

2.3.4 Food Sharing Through Meals

During my stay in Akulivik from mid-December 1990 to early January 1991, I observed 17 meals (lunches or suppers) held at the house of my host ("A" of Fig. 2.3). During this period, his household included himself, his wife, and their two daughters. At nearly every meal, several guests were present. Although they were sometimes invited by the wife, they consistently appeared only when the food was ready. These primary guests included members of A's extended family, friends, namesakes, immediate neighbors, and distant kinsmen (Kishigami 1992: 189). One case is shown as an example in Fig. 2.2 (Photo 2.3).

From late January to February and in July 1998, I stayed in the same household and again observed the pattern of meals. During this second observational period, the household consisted of A and his wife, their adopted son, and second oldest daughter. In January–February, I observed a total of 28 meals.

Photo 2.3 Food sharing through meal. Akulivik, PQ, Canada. December, 1990. By N. Kishigami

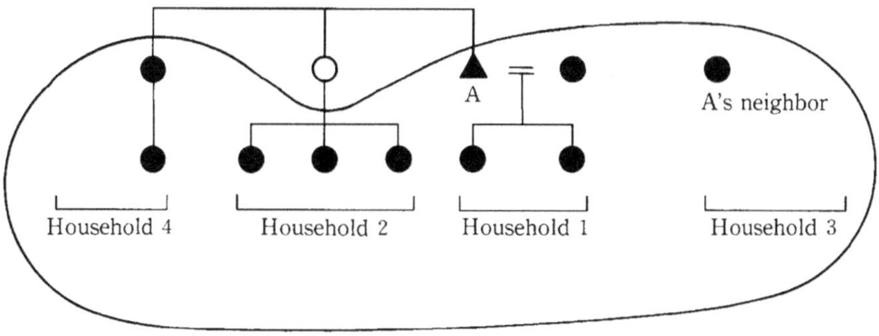

Fig. 2.2 Social relationships of the participants of A's supper at A's house. ▲ and ● Present

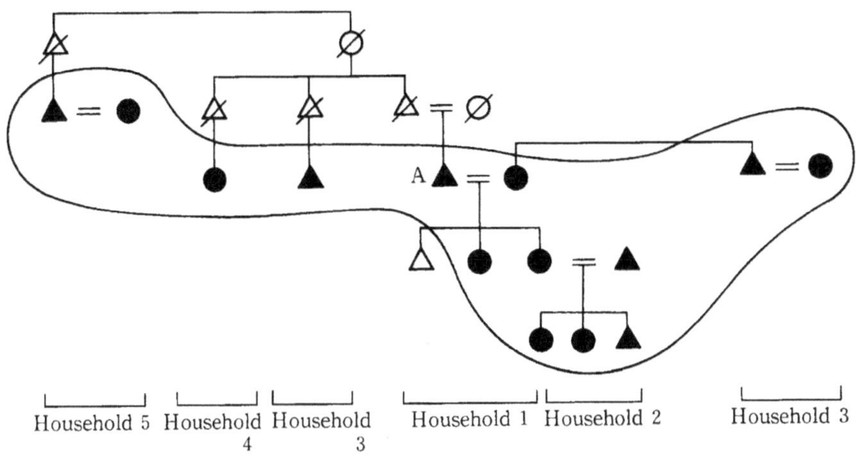

Fig. 2.3 Social relationships of the participants of lunch at A's house in February. ▲ and ● Present

There were only two occasions in which members of the household ate alone. At the other 26 events, members of A's extended family, along with more distant relatives and neighbors, were present and shared in the meals. One such meal is illustrated in Fig. 2.2.

I recorded another 34 meals at A's house in July. There were only four cases when household members partook of a meal without guests. At the other meals, the circle of participants was essentially identical to that present in January–February (see Fig. 2.4).

Of the 62 meals observed, 56 (90%) involved members who were A's kinsmen and/or unrelated but immediate neighbors. I conclude from those cases that a household (the principal main unit of residence today) was not the exclusive unit of food consumption in Akulivik. Generally, members from several households of the extended family attended meals. This was especially the case where the household head was an elder. In such a circumstance, it was typical for his/her sons, daughters,

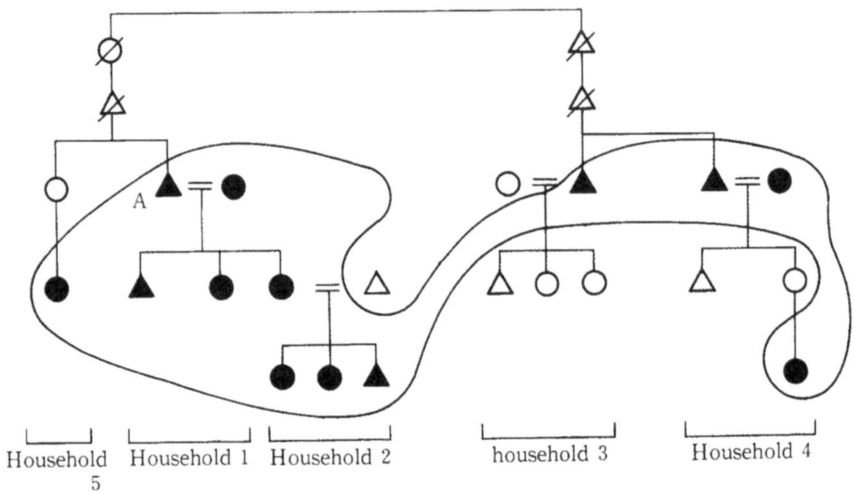

Fig. 2.4 Social relationships of the participants of lunch on July 11, 1998 at A's house. ▲ and ● Present

nephews, and nieces to give their catch to the elder and to later visit this central household for meals. Furthermore, distant kinsmen and other villagers might join these meals although less frequently than close kin.

2.4 New Food Sharing: Food Distribution via the Hunter Support Program and Through Local FM Radio Announcement

2.4.1 Hunter Support Program

The "James Bay and Northern Quebec Agreement" (1975) resulted in a "support program for Inuit beneficiaries for their hunting, fishing, and trapping activities" in the Nunavik region. The Inuit requested the government of Quebec to establish the program to maintain their vital hunting and fishing activities and to obtain country food. The program was legalized in 1982 through Bill 83 of the Quebec Provincial Government.

The aim of the program is "to favor, encourage and perpetuate the hunting, fishing and trapping activities of the Inuit as a way of life, and to guarantee the Inuit communities a supply of produce from such activities."[5] The primary use of the Hunter support program in Akulivik between 1990 and 2003 was to (1) employ the community boat for walrus hunting for a week from late September to early October, (2) use the boat for beluga whale hunting for a week in October, and (3) facilitate the free distribution of seal meat, caribou, ptarmigan, arctic char, and mussels bought from local hunters to villagers (Kishigami 2000).

Each village administered and used the program in their own ways under the provincial regulations. Thus, each village council could obtain a community hunting boat, acquired a communal cold storage house, or purchased meat and fish from local hunters with the program's funds to provide these free of charge to villagers in need in the 1990s.

The program budget is provided by the provincial government to the Kativik Regional Government of northern Quebec, which, after taking 15% of the budget for regional administration, redistributes the remaining amount to each village. The program allocation to Akulivik was approximately CA$115,000 in 1989, CA$139,000 in 1990, CA$142,000 in 1991, CA$150,000 in 1992, CA$182,000 in 1993, CA$135,000 in 1994, CA$208,000 in 1995, CA$523,000 in 1996 and CA$185,000 in 1997, CA$160,000 in 1998, CA$161,000 in 1999, CA$200,000 in 2000, CA$248,000 in 2001, and CA$207,000 in 2002. The annual community budget is subject to yearly agreements with the Kativik Regional Government. The amount is given to the community according to its beneficiary population, inflation rates, and proposed budget. The sums spent on the hunter support program have basically increased over the years due to the rapid population growth of the Inuit, with the increase especially pronounced between 1994 and 1996 due to the community's need as they hoped to repair their community boat.

2.4.2 Enactment and Use of the Hunter Support Program in Akulivik

In this section, I will describe how the Akulivik people used the hunter support program in 1997. The primary use of the program in the village was to (1) employ the community boat for walrus hunting for a week from late September to early October, (2) use the boat for beluga whale hunting for a week in October, and (3) facilitate the free distribution of seal meat, caribou, ptarmigan, arctic char, and mussels bought from local hunters to villagers.

In January and February 1997, the village councilors in charge bought five bearded seals and 10 ringed seals specifically for redistribution within the village through the program. The councilors purchased the meat of bearded and ringed seals for CA$2.5 per pound from local hunters and distributed this meat free of charge to the households of elders and widows. When there was sufficient meat, the council also gave meat to other households. Fifteen pounds of meat can provide several days of meals for the average household in Akulivik.

In March–April 1997, the village councilors purchased seals and caribou through the hunter support program.[6] They distributed the meat to the households of elders, widows, and others. In April, about 40 caribou were obtained from local hunters. On one occasion, the meat from 10–15 caribou was distributed to all the households in Akulivik.

From May to August, food was plentiful in Akulivik because of large harvests of Canada geese and snow geese. Arctic chars were abundant near the shore in July and August. The Akulivik Inuit caught a considerably large amount of game for themselves, which they shared with their kinsmen and other villagers. Thus, during this time, the councilors did not purchase any food through the program in the village.

For a week in late September–early October, the councilors, as part of the hunter support program, sent six hunters to Nottingham Island (more than 100 kms away from the village) for walrus hunting. Two of the hunters acted as captain and first mate of the community boat. The other four persons were selected by the mayor and councilors from the total group of hunters. A privately-owned boat accompanied the community boat on the walrus hunt. The councilors provided both boats with gasoline (200 gallons each) and CA$1000 worth of food and ammunition from the program budget. The program also paid CA$50 to CA$100 per day to each hunter for their service. The hunters were allowed to keep the walrus tusks for themselves, but could not take more meat than other villagers. Five walruses were caught, and the meat was equally divided and distributed to each household in the village, i.e., about 100 pounds each.

For about a week in October, a beluga whale hunting expedition was carried out using the community boat. Akulivik hunters used to hunt for Beluga whales in the

Photo 2.4 Equal distribution of beluga maktak to all Inuit households through the hunter support program. Akulivik, PQ, Canada October, 1999. By N. Kishigami

Richmond Gulf in the southeastern part of Hudson Bay. However, in 1994, they changed their whale hunting ground to a place further north near Ivujivik. In 1997, due to lack of program funds, the village was not able to send the community boat for beluga whale hunting. Instead, local hunters used two private boats. The councilors provided CA$500 worth of food and ammunition to each boat. Also, the Akulivik Co-Op paid CA$100 to each hunter who participated in the hunting expedition. After the hunt, 5% of the *muktuk* was shared by the hunters. The remaining *muktuk* and all the meat was equally divided and delivered to the 87 households in the village (Photo 2.4).

In November, many Inuit in Akulivik engaged in net fishing for arctic char at Kuuvik. In the last few years, the councilors have not organized and sent fishing groups to Kuuvik through the program. Rather, support funds were used to buy fish from the local Inuit and to distribute the fish to all the households. In 1997, the councilors purchased 2000 pounds of arctic char with CA$5000 (CA$2.50/lb) from the program. Each household was given 10 fish, supplemented by meat and fat of several caribou that were bought and distributed to several villagers that month.

In December, the councilors purchased 200 pounds of arctic char with the program funds for the community-wide Christmas feast. Usually, meat and fish purchased via the hunter support program were distributed among the households soon after their arrival at the village. However, some meat and fish were kept at the community storage. When any person needed the meat or fish, he/she could go to the storage to get some or ask a village official to deliver some to his/her home. Also, the community council and the church committee provided meat and fish to the villagers from the hunter support program at community feasts on several occasions, such as New Year's Day, Easter, Canada Day, and Christmas (Photo 2.5).

In addition to the above-mentioned activities, the program also supported the following activities: carrying out search and rescue operations (budget of CA$10,000), undertaking repairs to the community boat, making trails for snowmobiles and other four-wheel vehicles (all-terrain vehicles), deepening waterways for canoes, and organizing classes where elders taught hunting and survival skills to youngsters.

2.4.3 *Effects of the Hunter Support Program on Akulivik Society*

In this section, I will outline the program's activities and economic effects during a two-month period from February 17 to April 16, 1998 in Akulivik.

Table 2.1 shows the dates when meat was purchased through the program, the number of caribou bought, and the weight of seal meat. The age cohort of local hunters selling through the program, the volume of meat sales by each hunter, and the program income of each hunter are presented in Table 2.2.

Photo 2.5 Community feast at Christmas. Akulivik, PQ, Canada. December, 1990. By N. Kishigami

During the two-month period, the program obtained 132 caribou and 2749 pounds of bearded and ringed seal meat from the total number of 87 hunters for redistribution to needy villagers. The hunters earned a total amount of CA$20,590 from the program.

According to the mayor of Akulivik, if more than 10 caribou are available, they are divided and delivered to all the households. If fewer than 10 caribou are available, meat goes first to the households of elders, widows, and persons with full-time jobs who often help their father and brothers' subsistence activities by providing them with money or gasoline, and only then to other households in need of meat.

During these two months, there were six times when all the households in the village received caribou meat and nine other times when distribution was restricted to widows, elders, and full-time wage earners.

Assuming that one caribou can provide for five persons for two weeks, the hunter support program secured food for 9240 person/days (5 persons × 14 days × 132 caribou). This means a total of 23 days' food for the Akulivik population. In addition, there were 17 purchases totaling 2750 pounds of seal meat.

As I described earlier, the first focus of the hunter support program was widows, elders, and full-time wage employees who could not go hunting and fishing. These individuals/households benefited most frequently from the hunter support program, followed by villagers who were unsuccessful in their hunting and fishing expeditions and required occasional support. These two groups of Akulivik Inuit rated the program as being highly successful.

Table 2.1 The dates on which seal or caribou meat was purchased from hunters through the program, the number of caribou purchased, and the weight of seal meat purchased

Date/month (1998)	# of caribou	Weight of seal or fish (pounds)
17/02	12	459
18/02	6	366
23/02	2	223.8
26/02	0	115.4
27/02	10	0
02/03	14	0
05/03	0	124.6
06/03	0	194.6
09/03	21	40.8
10/03	2	229.8
11/03	0	34.8
13/03	2	0
17/03	0	20
18/03	8	0
19/03	21	0
20/03	4	260
23/03	0	36
24/03	12	118.8
25/03	8	72
26/03	0	140
27/03	0	213
03/04	3	0
13/04	3	0
15/04	4	0
16/04	0	100
Total	132	2748.6

On the other hand, the program was seen as problematic with regard to a third group. This group consisted of young Inuit in their 20s to mid-30s who only received meat from the program and did not contribute to food production. This group neither participated in hunting and fishing nor sought employment. Thus, they depended on transfer payments from the federal and provincial governments and drew heavily on food from the hunter support program and their kinsmen. As a result, they were seen as not fulfilling their responsibilities to their kinsmen and to the village. It was interesting to note that these young people did not hesitate to receive some meat and fish from their kinsmen or the hunter support program because they thought that being Inuit, they had the right to freely obtain them from other Inuit or the hunter support program. They did not give any food to other Inuit simply because they did

Table 2.2 The age cohort of local hunters selling through the program, volume of meat sales by each hunter, and the program income of each hunter

Hunter's ID#	Times to sell meat	Earned cash ($)	Age cohort ('s)
1	1	144	30
2	1	440	50
3	1	274.93	40
4	1	110	40
5	1	420	50
6	1	440	40
7	1	110	20
8	1	90	20
9	1	90	20
10	2	175.50	50
11	2	199.50	40
12	2	319.50	40
13	2	310	20
14	2	232	20
15	2	880	60
16	3	439.50	60
17	3	930	40
18	3	1240	30
19	3	1190	50
20	3	292.50	40
21	3	1300	50
22	4	1410	30
23	5	1808	50
24	6	1781.50	60
25	6	1680	70
26	8	1227	60
27	8	984	40
28	11	2071.50	40
Total	87 ($236.7 per time)	$20,589.43 ($735.34 per person)	

not have any, but they did not see that as a problem. The reality was that there was a discrepancy between the Inuit norms on food sharing and the young Inuit behavior toward and cognition about these practices.

2.4.4 Impact of the Hunter Support Program on Hunters in Akulivik

One of the reasons that the hunter support program was established by the Nunavik Inuit was to provide cash income to full-time hunters in their middle and old age who could not participate in wage labor due to their lack of English language competency and education. During the two-month period discussed here, 28 hunters sold their catch to the program, receiving almost $20,600. In terms of age cohorts, there were four hunters in their 20s, four in their 30s, nine in their 40s, six in their 50s, four in their 60s, and one in his 70s.

Those 28 hunters sold meat 87 times. On average, a hunter received about CA$237 each time. However, there was considerable variation, from CA$90 to CA$2,070, in any hunter's total income during the period. On average, a hunter's income was about CA$735 during this period.

Tables 2.3 and 2.4 show that higher the age of the hunters, the greater the income from the program and the higher the frequency of sales to the program. Although no individual's income from the program was sufficient to meet all cash needs, the hunter support program offered one of the few secure income sources for hunters aged over 40. However, it was also apparent that full-time hunters could not be

Table 2.3 Income from the hunter support program and age cohort of hunters who sold meat to the program

Income age cohort	Less than $99	$100–$499	$500–$999	$1000–$1499	$1500–$1999	$2000–$2500
20's	2	2				
30's		2		2		
40's		6	2			1
50's		3		2	1	
60's		1	1	1	1	
70's					1	
Total	2	14	3	5	3	1

Table 2.4 Frequency of selling meat by each hunter to the program and age cohort of the hunters

	Once	2 times	3 times	4 times	5 times	6 times	8 times	11 times
20's	3	1						
30's	1	1	1	1				
40's	3	2	2				1	1
50's	2	1	2		1			
60's		1	1			1	1	
70's						1		

economically independent solely by selling their catches to the program. The village used the program budget to pay money to hunters for their procurement of animals and/or fish or to buy animals or fish that they had caught. This was a kind of economic exchange between the village and hunters.

2.4.5 Food Sharing Through Local FM Radio Announcement

When a hunter caught a large animal, such as a beluga whale, walrus, or bearded seal, or a large number of fish or birds, he or his wife made an announcement of food sharing through the local FM radio broadcast in Akulivik. He or she invited the villagers to come to his/her house to pick up some meat or fish if they wanted a share. There were several productive hunters and/or fishermen who sometimes got a big animal or large quantity of fish or birds. They sometimes shared their game with other villagers in this way. While they thought they were socially obliged to provide food to other Inuit or the whole community, they also experienced joy in sharing their catches with others, which they aimed for in daily life.

As this sharing implied a public acknowledgment of how the hunter was a good and generous person, he earned a good reputation and respect from other villagers, which increased his social influence in the community regardless of whether this was his intention or not. However, as far as my observation was concerned, this kind of sharing practices did not always give political power to the hunter concerned in Akulivik, where devout Christian elders and heads of extended families wielded significantly greater political power than the hunters, which was reflected in who the mayor and community councilors were (cf. Ready 2016). However, I do not necessarily deny the possibility that some hunters such as mayoral candidates used their generous sharing practices to gain popularity in the village.

There was another way of sharing through the use of the local FM radio. When a person found that his/her relatives did not have food or other things that they needed, he/she asked other villagers to give such things to them. There were direct requests for food and other resources made to other villagers through FM radio broadcasts. For example, when a young woman was out of powdered milk or disposable diapers for her baby, cigarettes, or flour, she made a telephone call to the local FM station to request other villagers to give her the necessary goods. If somebody in the village had any to spare, he/she would provide her with it free of charge.

2.5 Discussion

Many Inuit elders in Akulivik expressed the opinion that sharing food was the only way for them to survive in the 1980s–1990s. The contemporary Inuit of Akulivik responded to economic situations by a combination of three means of food aid: (1) customary food sharing practices, (2) food sharing through the hunter support program, and (3) food sharing by use of local FM broadcasts. While (1) was traditional, (2) and (3) were new forms.

Akulivik hunters still shared game first with close kinsmen, such as parents, grandparents, siblings, children, uncles, aunts, and cousins, and then with other villagers. Research results on household activities from 1990 (Kishigami 1992) and 1998 indicate that food sharing was most frequently practiced among households within the extended family and then among physically close neighboring households in the community. Several hunters sometimes also gave food to elders, sick persons, and widows, in addition to namesake persons, symbolic midwives, or friends. Further, indirect food sharing was practiced through participation in meals at the homes of successful hunters. Some persons in need asked for food from successful hunters, which might be characterized as a kind of sharing based on weak demand.

The hunting and fishing activities of the Inuit were not so extensive in Akulivik as to always meet the demand for food in the 1990s. However, this situation was slightly improved by the Inuit's use of the hunter support program, which was initiated in 1983 to provide country food to needy households or/and all the villagers in the community. The sharing system through the hunter support program aimed as much as possible to provide country food to the needy or/and all the households in the community.

In the winter and fall seasons, when lack of food was most common, the hunter support program provided country food to the Akulivik Inuit in need and sometimes to all the households. In the summer season, when there was abundant food, hunting and sharing was carried out extensively within each extended family. Also, food giving across the village was done voluntarily or by request through radio announcements. In these ways, food and other necessities reached the Inuit in need and also others. Many or all Inuit households received some meat and fish through the hunter support program and existing food sharing networks. Once food was obtained by the Inuit, they shared it with their kinsmen and neighbors again and again.

In most cases, the Inuit who offered food to others did not expect to be paid back or reciprocated by the receivers. For the Inuit, giving food to needy persons was obligatory (ex. Riches 1981). They simply gave their food to others or received food from others. Each transaction was completed by the single action of giving or receiving. The transaction tended to be not reciprocal between two persons except between close kinsmen. Thus, I argue that for the most part, Inuit food sharing did not constitute a form of reciprocal exchange but was a one-way transfer or distribution of food. It should be emphasized that because Inuit food sharing was practiced mostly among persons in close social relationships, food seemed to be intentionally and reciprocally exchanged between them over a long period.

I will examine some social functions and effects of these sharing practices.

First, the sharing practices contributed to people's survival because the have-nots could obtain food from either of the sharing practices. Although customary food sharing was conducted primarily within each extended family and also aimed at elders, widows, etc., food sharing through the hunter support program and the use of local FM broadcasts was carried out across several or all extended families in Akulivik. The food sharing practices functioned as an economic and nutritional insurance among the Inuit. However, as the Inuit shared food mostly with their extended family members, all the Inuit or Inuit households did not necessarily receive equal amount of food. If there are more hunters in an extended family consisting of several households, those households have a higher possibility of receiving more food than households of an extended family with fewer hunters. Thus, there were some economic and nutritional differences among households belonging to or relating to different extended families. However, the situations were not overcome but supplemented through food sharing via the hunter support program and the use of local FM broadcasts across the extended families in the village.

Second, the food sharing practices had positive social effects in terms of reproducing the distinctive Inuit social relationships in Akulivik. Food sharing constituted a part of the Inuit social subsistence system (Wenzel et al. 2000; Hovelsrud-Broda 2000; Kishigami 2004; Wenzel 1995, 2000, 2013, 2019). Among the Inuit, the hunting, sharing, and consumption of animals were socially organized practices carried out based on Inuit social relations (ex. Loaner 1980; Ellanna and Sherrod 1984; Langdon 1991; Wenzel 1981, 1991). Within this system, the sharing of food with other Inuit maintained or reproduced the social relations between givers and receivers of food. As long as food sharing was practiced by the Inuit on the basis of kinship and other social relationships such as namesake partners, midwife and hunting partnerships, and friendships, those social relationships were activated and maintained.

Third, the Inuit sharing practices were closely linked to an Inuit hunters' worldview, which was shaped by the relationship between humans and animals. The Inuit believed that because animals gave themselves up to hunters, it was incumbent upon the hunters to in turn give them to others (Bodenhorn 2000: 44–47; Fienup-Riordan 1983:346; Nuttall 1991: 219; Omura 2021: 67; Stairs and Wenzel 1992). While the worldview of the relationship between the Inuit and animals encouraged the Inuit to continue sharing food with others, the practices reproduced the worldview.

Fourth, hunters gained social reputation and respect from other people through their generous food sharing practices. As the Inuit were taught from childhood that food sharing was one of their regular behaviors to be undertaken without any question, they believed that they should not show off their generosity. Thus, they shared food with others daily in a humble way or as a matter of course. However, frequent givers of food were able to acquire good social reputation and respect from receivers. Sometimes, this reputation and respect led to political power in the community. Also, many hunters believed that they should feed other people in the community as well as experienced joy in giving food to others to help them. It seemed food sharing was a goal in itself for several good hunters in Akulivik.

Fifth, Inuit sharing practices were associated with familial, village, and Inuit identities. As I pointed out that most food was customarily shared with extended family members, this kind of food sharing brought about the "we-feeling" within each extended family. The food supplied by the hunter support program facilitated food sharing at the village level. Although this communal aid seems to have been an emergent phenomenon in the modern context, it was in reality a carryover from the small community (camp group) situation that characterized the Nunavik Inuit before sedentarization. The Inuit regarded resource sharing for mutual help as one of their traditional social characteristics. The sharing practices at the village level across all extended families or households maintained and strengthened their local group identity as Akulivik residents and, furthermore, a feeling of being real Inuit.

While I discussed the positive functional features of the contemporary food sharing practices, they had another aspect. I should admit that the practices of the hunter support program might be causing social changes in the Akulivik society. Through the program, the Inuit began to receive meat and fish from unspecified individuals, thus weakening reciprocal obligations or responsibilities to food givers. Furthermore, some young men and wage workers who did not engage in hunting and fishing activities became net receivers of meat and fish through the program. At the time, there remained strong kinship norms on food sharing and mutual help practices among the Akulivik Inuit. Wage workers and some young Inuit without subsistence activities still received some meat and fish from their relatives and via the hunter support program without stigma. However, the social differentiation of the Inuit into two categories, harvesters (givers) and net-receivers of country food, might cause some alterations in existing social relationships in the long run.

The hunter support program was highly valued by the Akulivik Inuit in terms of its overall economic and social effects. However, the implementation of the program as well as the decline in fur trading and hunting/fishing activities in general might have some devastating effects on food sharing networks of young Inuit within and across extended families (cf. Collings, Marten, Pearce and Young 2016). In order to keep the program functioning effectively in socio-economic terms, efforts should be taken to involve the young Inuit who do not engage in subsistence activities in the hunter support program and other indigenous aid systems.

2.6 Conclusion

In the case of the Akulivik Inuit, there were three kinds of food sharing practices: (1) customary food sharing practices, (2) food sharing through the Hunter Support Program, and (3) food sharing by use of local FM announcements in the 1990s. Among them, sharing through the hunter support program and local FM announcements supplemented customary country food distribution. These sharing practices have made the reproduction of distinctive Inuit social relationships possible.

One common characteristic of these practices, however, was that they did not necessarily amount to a reciprocal exchange of food between specific persons or groups, but instead represented a one-way food transfer from one person to another/others or a redistribution of game from the community to its members.

Food sharing was one of the social characteristics of small-scale societies such as hunting-gathering societies (Altman 1987; Damas 1972; Lee 1979; Sillander 2019; Woodburn 1981). The existence of such sharing practices among the Inuit has been primarily explained in terms of their functions of insurance for survival and social cohesion (Collings et al. 1998; Kelly 1995: 161–181; Wenzel 1995). The Akulivik Inuit themselves recognized those functions. In this chapter, I emphasize the nutritional, economic, social, and cultural effects of Inuit food sharing practices among the Akulivik Inuit.

While hunting and fishing activities have drastically declined since the 1980s, the hunter support program, implemented in 1983, has led to the distribution of meat and fish to all the Inuit households in Akulivik. The fish and meat were then shared on the basis of social relationships such as kinships, co-hunting partnerships, and other relationships. Food sharing promoted by the hunter support program contributed to the maintenance of Inuit social relationships despite the rapidly globalizing politico-economic circumstances affecting the Inuit society.

To the contemporary Inuit, food sharing practices were, generally speaking, economically, socially, and culturally significant because they correlated with several aspects of the Inuit value system. Changes in these practices might bring about fundamental alternations of a worldview on the special relationship between the Inuit and animals.

In the current socio-economic context of Nunavik, the existing sharing system has been in decline due to the growing prevalence of the globalizing cash economy, diversifying Inuit lifestyles, rapid population increase, and implementation of wild resource management of beluga whale and walrus. I argue that Inuit food sharing is a reliable indicator of social change in some cases and of continuity in others, as it is organized and practiced among specific social relationships and as it reflects the Inuit worldview.

Notes

1. All the Inuit of Ivujivik and Puvirnituq and one half of the Salluit Inuit in Nunavik have disapproved of the agreement.
2. Around the beginning of the twenty-first century, country food supplied 16% of the total calories in comparison with 84% supplied by store-bought foods (Blanchat and Rochatte 2008: III).
3. An extended family functioned as an economic unit. There was mutual help among households of kinsmen from an extended family. There were three large hunting boats in Akulivik, two of which were privately owned by the local Inuit. Each boat was purchased, owned, managed, and used by an elder in cooperation with his brothers, sons, and nephews. They collectively provided the money and labor to run the boat for hunting and other purposes. In this case, a large extended family was the social unit managing the boat.
4. A symbolic midwife is a person who puts the first clothing on a newborn baby while whispering his/her wish to the baby on the east coast of the Hudson Bay in Nunavik. The midwife and the baby establish a special bond between them. The former teaches the latter in his/her childhood and gives many presents. On the other hand, the latter has to give all of his/her first catch or things made to the former. See Guemple (1969).

5. The Cree established their own hunter support program that was distinct from the Inuit's. Regarding the Cree hunter support program, see Feit (1982, 1991), Scott (1984), and Scott and Feit (1992).
6. The purchase price of a whole female or young male caribou was CA$110 and that of an adult male CA$100.

Chapter 3
Food Sharing Among the Iñupiat in Utqiaġvik, Alaska, USA

3.1 Introduction

The Iñupiat live in Northwest Alaska. Those living along the coast began to actively hunt bowhead whales around the tenth century. Since then, whaling has formed the socio-economic basis of their society. Thus, they have more than a 1000-year history of bowhead hunting in the region (Savelle 2005). Even now, they continue to maintain an identity as a whaling people or "people of the whales" (Sakakibara 2010).

I conducted research on the Iñupiat's bowhead whale hunting and the sharing and distribution of bowhead whales in Utqiaġvik (formerly known as Barrow) in Alaska from 2006 to 2013. The bowhead whale (*Balaena mysticetus*)—a baleen whale—grows to be approximately 15 m long and 70 tons heavy. Hunters in Utqiaġvik prefer to hunt small whales that are approximately 10 m in length. Even a single small whale brings approximately 2200 kg of meat and 1500 kg of *maktak* (skin parts containing fat) to a village. These products are divided according to specific rules. They are subsequently shared extensively throughout the community. In addition to whale meat and *maktak*, the meat of polar bears, beluga whales, walruses, bearded seals, ringed seals, caribou, fish such as salmon, and birds such as snow geese and ducks are also frequently shared among hunters and/or among hunters and other villagers in Utqiaġvik.

Utqiaġvik is located at 71°29′N latitude and 156°79′W longitude and is the northernmost village in the United States. According to the 2010 census, the town's total population was 4974, of which 65.0% were Iñupiat, 16.0% were Euro-American, and 19.0% were others, which included immigrants from the Philippines, Thailand, Korea, etc. (North Slope Borough 2012). The economic system of Barrow relies on market and subsistence economies, which is akin to many other villages in the Arctic. The majority of Iñupiat adults engage in wage labor throughout the year and carry out their hunting and fishing activities on weekends, holidays, and between work shifts. Many of the Iñupiat hunters engage in bowhead whale hunts for a few weeks in spring and fall respectively, which is when bowhead whales migrate offshore near Barrow. This whaling is one of the greatest concerns among the Iñupiat in Barrow.

N. Kishigami, *Food Sharing in Human Societies*, Trust 4,
https://doi.org/10.1007/978-981-16-7810-3_3

This chapter primarily describes and discusses the bowhead whale sharing among the Iñupiat in Utqiaġvik. It first provides a holistic picture of food sharing in traditional Iñupiat society. It then describes the sharing of whale products and other food, and examines food sharing in contemporary Iñupiat society.

3.2 Food Sharing Practices in Traditional Iñupiat Society

The Iñupiat are hunter–gatherers whose mother tongue is the Iñupiaq language. They live in the northwestern region of Alaska. According to Burch (1980, 1988), there were 26 regional societies of the Iñupiat in the northwestern region of Alaska during the period ranging from 1816 to 1881, around when the Iñupiat first came into contact with Europeans. A regional society was an independent societal entity consisting of several bands and was equivalent to a nation. The societies from the 1816–1881 period examined in this study are expediently termed traditional societies.

Burch (1980) published a paper pertaining to the property (assets) and transfers of property among the Iñupiat in northwest Alaska. He distinguished five different types of property among the Iñupiat based on ownership: societal, local family, domestic family, conjugal family, and individual.

The property of each regional society refers to the land that the people occupy as their territory. Each regional society includes multiple band groups, which constitute the everyday social units of Iñupiat society. Each band group is comprised of several family households that are mutually connected through multiple kin relationships. Burch termed the band group that comprises of multiple family households as a "local family." Since a "local family" is equivalent to an "extended family"—the term used in anthropology—this chapter uses the latter term. The property of an extended family included a communal house, called *qazgi*, and food. Food was kept and managed by the head of an extended family. Burch termed a family whose members occupied a single dwelling as a "domestic family." The property of a domestic family included food, a house, tents, *umiaq* (a large boat made of sealskin), and dogs and sleds. In addition to the two family types, Burch designated a family comprised by one married couple as a "conjugal family." The property of a conjugal family included bedding and fur. There were individual possessions too. The property of individuals included clothing, tools, hunting instruments, kayaks, household items, amulets, charm songs, and raw materials. Therefore, food belonged to an extended family (local family) and/or a domestic family in traditional Iñupiat society.

In the following section, I will draw from Burch's (1988) study to outline food sharing practices in traditional Iñupiat society as a baseline for this study.

3.2.1 Sharing Practices in Traditional Iñupiat Society

The Iñupiat established several methods for the transfer of goods, such as exchange, trade, barter, sharing, redistribution, loans, inheritance, and theft (Burch 1988: 102–106). They practiced food sharing and redistribution daily. Barter was practiced among different regional societies, extended families, and individuals. This section provides an overview of allocation systems for goods other than food.

Direct exchange of similar goods between two parties was called *simmiq*. One such example is the exchange of large boots for small boots. The method of acquiring goods that is equivalent to our purchase of merchandise was called *tauqsiq*. The exchange of the seal blubber that one might have for caribou (wild reindeer) fur that one might want is an example of such an exchange. An exchange was called *tunilaq* when initiated by the other party. The exchange of goods between trading partners belonging to different regional societies was called *nuviq*.

There were two types of loans in traditional Iñupiat society. The lending and borrowing of clothing with the assumption of utilizing them completely or to a significant extent was called *aturakasak*. However, the loan of goods with the assumption of returning them after use was called *ataqsi*. Upon the demise of the head of a household, possessions such as *umiaq*, sleds, songs, and amulets were inherited by the successor. Inheritance was called *kinguvaanaqtuq*. Theft was called *tiglik*.

In the nineteenth century, as many as 2000 people would gather at large trade fairs held in Sisualik and the Colville River Delta on the north side of Kotzebue Sound in Alaska for several weeks every year to exchange goods, dance, hold competitions, and eat together. Trade fairs centered on barter among trade partners. The participants subsequently also exchanged goods with parties other than their trade partners. Through such trade, tobacco and metals from Russia, reindeer hide from Siberia, and glass beads had been distributed in northwest Alaska even before the Iñupiat came in contact with Europeans (Burch 1988: 105–106). The Messenger Feast that was held in late fall or early winter was similar to the trade fair, except for its smaller scale. People gathered for several weeks and exchanged goods, danced, held competitions, and ate together (Burch 1988: 106).

The next section introduces the daily food sharing practices of the Iñupiat of northwest Alaska.

3.2.2 Meat Sharing After Hunting

When an individual hunter hunted a polar bear, ringed seal, or any other animal, the game belonged to the hunter. However, the game would be divided into several parts, and edible parts such as meat would be taken to the wife of the head of the extended family or domestic family. It thereby became shared property of one of the families. The fur of the game would become the property of the hunter and his wife (Burch 1988: 102).

Game from group hunting would be divided either equally or unequally among the participants. Caribou, bearded seals, walruses, and a lot of fish caught by a group would be mostly divided equally among the participants. This form of sharing was called *aviktuuzaaq* (Burch 1998: 102).

Whale meat obtained from a bowhead whale hunt was divided unequally. When a bowhead whale or gray whale was hunted, its edible parts were divided among the hunters in accordance with specific rules. Despite not pariticpating in the hunt, the people who helped butcher and process the whale also received a share. This form of sharing was called *ningiq*. Owners of large boats and the heads of extended families would, in some cases, received a greater amount of meat than others. This form of sharing was called *pillyuk*. In other cases, elders in a village or whaling group received a large amount of the best part of the hunted animal. This form of sharing was called *umiqtuat* (Burch 1998: 102).

Game was given also to people who happened to be at the place where the fish or animals were cut and those who watched the activity. This form of sharing was called *pikszak* (Burch 1998: 103).

In domestic and extended families, the provided food was consumed and there was expectation of reciprocity. This form of sharing—called *pigziaq*—was unrestrcitedly practiced among close family members. Successful hunters also distributed their game to elders, people who were unable to hunt because of illness or injury, and widows in other domestic and extended families. This was supplied to them free of charge. This form of sharing was called *aiccuq* (Burch 1998: 104–105).

3.2.3 Basic Characteristics of Food Sharing in Traditional Iñupiat Society

According to Burch's research, a village or band group in traditional Iñupiat society often consisted of a single local family (extended family). Hence, if necessary, food was pooled at an extended or domestic family and redistributed to its members by the head of the family or his wife. Within a single family group, sharing based on generalized reciprocity was commonly conducted.

Successful hunters distributed their game to elderly people, sick people, and widows belonging to families other than theirs. Furthermore, food was shared beyond the boundaries of regional societies through communal feasts at trade fairs and at the Messenger Feast.

During the nineteenth century, the Iñupiat participated in fur trade and commercial whaling with Europeans and Euro-Americans through contact with them. This chapter primarily concerns the period ranging from the end of the twentieth century to the beginning of the twenty-first century. During this period, the Iñupiat lived sedentary lives in several villages, each of which was composed of multiple extended families.

3.3 Whaling-Related Food Sharing

3.3.1 *Whaling in Utqiaġvik*

Whaling conducted by the current Indigenous peope in Alaska is categorized as Aboriginal Subsistence Whaling and is regulated by the International Whaling Commission (IWC). Aboriginal Subsistence Whaling is a type of whaling approved by the IWC (International Whaling Commission) in view of historical, dietetic, and cultural needs. It is defined as "whaling for purposes of local aboriginal consumption conducted by or on behalf of aboriginal indigenous or native peoples who share strong community, familial, social, and cultural ties related to a continuing traditional dependence on whaling and on the use of whales" (IWC 1981).

The IWC had been reviewing the whaling quota for the Arctic Ocean area off the coast of Alaska every five or six years from 1998–2018. The hunting of 280 whales was permitted between 2008 and 2012. Permission for the hunting of 25 whales out of the total of 280 is assigned to the indigenous peoples of Russia. Therefore, the Iñupiat and Yupiit in Alaska are allowed to hunt upto 255 whales (51 per year).

Currently, the Iñupiat and Yupiit are the only people who hunt whales in the state of Alaska. Iñupiat villages include Little Diomede, Wales, Kivalina, Point Hope, Point Lay, Wainwright, Utqiaġvik, Nuiqsut, and Kaktovik. Yupiit villages include Gambell and Savoonga on Saint Lawrence Island (Map. 3.1 Indigenous whaling villages in Alaska). Presently, several indigenous people of Alaska use whales primarily as food. They also use the baleen and bones of whales as materials for crafts. However, such use is rather rare now.[1]

Utqiaġvik, the community I researched, is located at 71°29′N latitude and 156°79′W longitude. It is the northernmost village in the United States and it faces the Chukchi Sea. The Iñupiat of Utqiaġvik hunt whales that migrate in the sea near the village in spring and fall. The spring hunting season approximately ranges from late April to mid-May. The fall hunting season extends from around the end of September to mid-October. This community currently has approximately 50 whaling captains (groups) and more than 300 whaling boat crew members (hunters) (Photo 3.1).

In Utqiaġvik, 29 whales (16 in spring and 13 in fall) were caught in 2005, 22 whales (3 in spring and 19 in fall) were caught in 2006, 20 whales (13 in spring and 7 in fall) were caught in 2007, 21 whales (9 in spring and 12 in fall) were caught in 2008, 19 whales (4 in spring, of which one could not be landed, and 15 in fall) were caught in 2009, and 22 whales (14 in spring and 8 in fall) were caught in 2010 (Suydam et al. 2006, 2007, 2008, 2009, 2010, 2011). Approximately 22 whales were caught per year on average during this period. These whales were approximately 10–11 m in length on average.[2]

Aboriginal subsistence whaling conducted by Indigenous peoples such as the Iñupiat and Yupiit are approved by the IWC when certain conditions are met. One such condition is that whaling needs to be a part of their tradition and subsistence even today, and is not a for-profit activity in principle (Hamaguchi 2002: 23–26; Iwasaki 2011: 213–214). Therefore, the Iñupiat and Yupiit do not sell whale meat

Map 3.1 Indigenous whaling villages in Alaska. (*Source* Alaska Eskimo Whaling Commission: http://www.aewc-alaska.org/whaling-villages.html)

and *maktak* after a successful hunt. This signifies that whaling is not a source of income for them. The current Iñupiat whaling is funded through the savings from the wage income of the captains, their wives, and other family members; dividends given by Indigenous corporations and Alaska's state government, and other sources.[3]

In Utqiaġvik, a total of approximately US$10,000 to US$30,000 are required for a single whaler group to conduct whaling activity during spring. Approximately US$5000 to US$10,000 are required as expenses for fall whaling. Boat maintenance, outboard motors, and other devices needed for whaling necessitate an expenditure of more than US$50,000 (Kishigami 2009a: see Table 5; Kishigami 2013a, b, c). Understanding this situation is a precondition for one to comprehend the contemporary whaling activity of the Iñupiat people.[4]

3.3.2 *Butchering and Sharing of Whales by the Iñupiat in Utqiaġvik*

The ways in which the Iñupiat butcher and share whales are described in records left particularly in the village of Point Hope (Foote 1992; Gamo 1964: 16–17;

Photo 3.1 Spring whale hunt. Barrow (Utqiaġvik), Alaska, USA. May, 2010. By N. Kishigami

Rainey 1947; VanStone 1962: 48–52; Worl 1980: 317–320). Burch (2006: 160–165) described the butchering and sharing of whales in the region of Wales around the year 1800. Additionally, George (1981: 789–803) published a report on whale butchering and sharing in Utqiaġvik in the latter half of the 1970s. Comparisons of these descriptions reveal that the sharing methods changed over time and indicate that sharing methods varied across villages.

This section first explains the Iñupiat's rules for primary distribution after the butchering of whales that are currently considered standard in Utqiaġvik. It subsequently describes the diachronic changes and synchronic variation in whale sharing. The standard sharing rules presented here were submitted for educational use to the North Slope Borough School District School Board of Education by the Barrow Whaling Captains Association (North Slope Borough School District 2002).

3.3.3 Whaling and Standard Sharing of Whales in Utqiaġvik

Whales that have been captured are butchered in line with specific rules. The methods of butchering and meat sharing vary across villages. Some differences even exist between captains within the same village. The following introduces the rules that are currently considered most common in Utqiaġvik (Fig. 3.1).

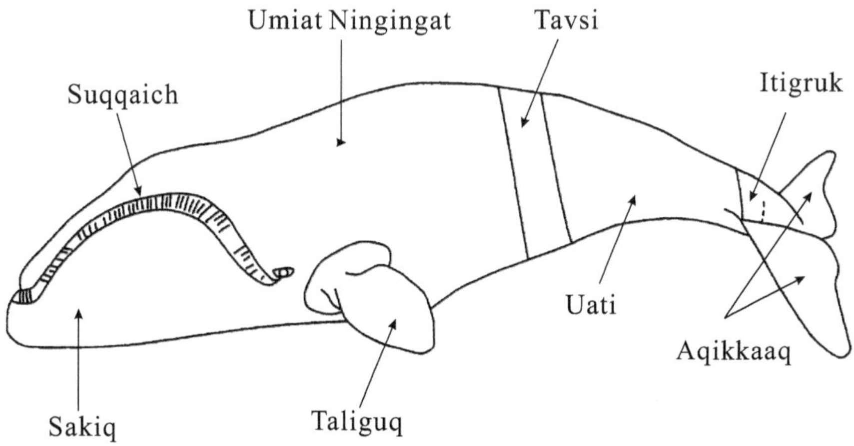

Fig. 3.1 Names of the various parts of bowhead whales that are divided (North Slope Borough School District 2002)

(1) *Tavsi*: Meat and *maktak* that are 30 cm in width are divided into two. Half goes to the successful crew. This half is divided as personal shares among the crew, under the supervision of the whaling captain: The other half is cooked and served to the public at the captain's house the day after the successful hunt.

(2) *Uati*: This is kept in the captain's ice cellar; it is served at feasts such as Nalukataq, Christmas, and Thanksgiving.

(3) *Itiġruk*: This part is kept in the captain's ice cellar; it is given to the visitors at Nalukataq.

(4) *Aqikkaak*: This part is kept in the captain's ice cellar; it is served at feasts such as Nalukataq, Christmas, and Thanksgiving.

(5) *Umiat Ningingat* or *Ningik*: All whaling crews who participate in the butchering of the whale share this part equally. It is noteworthy that this part is equally divided among all the whaling crews that registered during the first spring hunt.

(6) *Suqqaich*: Half goes to all the crews who helped with the towing. The rest goes to the successful crew.

(7) *Sakiq*: One side goes to the whaling captain; the other is divided among the crews who assisted in towing the whale.

(8) *Taliġuq*: One side goes to the harpooner; the other is used as "taquun" by all whaling crews at the butchering site.

(9) *Utchik* (tongue): Half is divided among all crews at the butchering site. Approximately one quarter is served at the successful captain's feast; the remainder is served at Nalukataq.

(10) *Uumman* (heart), *Ingaluaq* (intestine) and *Taqtu* (kidney): Half is served at the captain's feast. The remainder is served at Nalukataq.

Photo 3.2 Butchering a bowhead whale. Barrow (Utqiaġvik), Alaska, USA. May, 2010. By N. Kishigami

Once the butchering and the sharing of a whale are completed, the captain gives a signal. After this signal, anyone is free to take the remaining meat and any other parts of the whale. This distribution is called *pilaniaq*. This distribution system shows that meat and other parts of a whale are not owned exclusively by the crew that caught it. In reality, these items are shared, in accordance with the established rules, with other whaling crews and people who have helped butcher and/or tow the whale (Photos 3.2 and 3.3).

3.3.4 Synchronic Variation in Whale Sharing in Utqiaġvik Village

The following section introduces sharing methods of two young captains (mid-40 s) and an older captain (early 60 s), while comparing them to the standard sharing methods.

Case 1. H.A. (interviewed on August 22, 2010)

Photo 3.3 Sharing the meat and other parts among several whaling groups that helped butchering a whale. Barrow (Utqiaġvik), Alaska, USA. May, 2010. By N. Kishigami

H.A., who is in his mid-40s, has been a captain for approximately five years. The whaling group headed by H.A. captured a whale in May 2010. They divided the whale at that time as per the following the method.

(1) *Tavsi*: Same as the standard sharing procedure. Captain H.A. equally distributed half of it to the members of his whaling group. The remaining half was distributed to the people who prepared and cleaned up after the celebration feast at his home, which comprised approximately 30 individuals in all.
(2) *Uati*: Same as the standard sharing procedure. Approximately half of the part was used for the Nalukataq feast, whereas the remaining two quarters were used for Thanksgiving and Christmas, respectively.
(3) *Itiġruk*: used in the same way as the *Uati*.
(4) *Aqikkaak*: used in the same way as the *Uati*.
(5) *Umiat Ningingat*: All whaling crews, which include the captain's own whaling group who participated in the butchering of the whale and the others who assisted in it, shared this part. For instance, if there are 15 total whaling groups and five individuals who took part in the butchering, the five individuals are deemed a group and the catch is equally divided into 16 shares.[5] Each whaling group receives an equal share of whale parts, which is redistributed within the group at the discretion of their respective captains. Captain H.A. equally

distributed the catch among eight individuals, which included his four crew members, two fuel suppliers, the lender of whaling tools (darting guns), and himself. In principle, this is the same as the standard sharing procedure, but the fact that his own whaling group also took a share is a divergence from the norm.

(6) *Suqqaich*: Same as the standard sharing procedure.

(7) *Sakiq*: Same as the standard sharing procedure.

(8) *Taliġuq*: While giving one pectoral fin to a harpooner who killed the whale is the same as the standard sharing procedure, the captain took the other pectoral fin instead of giving it to people who took part in butchering. He let it mature and ferment and served it at the Nalukataq festival.

(9) *Utchik*: Captain H.A. did not provide half of the whale's tongue to all the whaling groups that engaged in the butchering, but instead served it at the captain's feast. The other half was served at the Nalukataq festival.

(10) *Uumman* (heart), *Ingaluaq* (small intestine), and *Taqtu* (kidneys): Same as the standard sharing procedure.

Additionally, akin to the standard sharing procedure, the captain gave a signal after the butchering and distribution were completed and allowed anyone to freely take the remaining meat and fat.

As described above, methods that differed slightly from the standard practice was apparent in the distribution of (5), (8), and (9). This result reflects the decisions made at the discretion of the captain.

Case 2. G.B. (interviewed on August 26, 2010).

G.B. became a captain of a whaling group that had one of the highest success rates in his village—succeeding to the position from his late father. Despite only being in his late 40 s, he is an experienced hunter. He captured a whale on May 5, 2010. He shared the catch in the following manner.

(1) *Tavsi*: Same as the standard sharing procedure. However, the captain, who owns the boat, took one more share.

(2) *Uati*: Same as the standard sharing procedure.

(3) *Itiġruk*: Same as standard distribution procedure.

(4) *Aqikkaak*: Considered to be a part of the *Uati* and was shared in a similar manner.

(5) *Umiat Ningingat*: Divided equally amongst all whaling groups and groups that participated in the butchering. However, unlike the standard sharing procedure, the whaling group that captured the whale also took part in the butchering and received a share.

(6) *Suqqaich*: A half of the baleen was given to the group that caught the whale. The other half was shared equally among groups that participated in the towing of the whale. In the latter case, the group that captured the whale took part in the towing and was given a share.

(7) *Sakiq*: One side of a whale from the mouth to the jaw was given to the group that caught the whale. The other half was divided equally among the whaling

crews who participated in the towing of the whale. In the latter case, the group that hunted the whale would also be given a share if it took part in the towing.[6]

(8) *Taliġuq*: Same as the standard sharing procedure.

(9) *Utchik:* Half of the tongue was given to all whaling groups that engaged in the butchering. The other half was divided equally into three shares, one of which was given to the members of his group that caught the whale, whereas another was served at the captain's feast. The remainder was served at Nalukataq.

(10) *Uumman* (heart), *Ingaluaq* (small intestine), and *Taqtu* (kidneys): Served at the captain's feast and Nalukataq as in the standard distribution procedure, but the shares were determined by the captain. Captain G.B. used a greater share for the captain's feasts than for Nalukataq.

(11) Eardrums: One was taken by the captain, whereas the other was taken by the harpooner.[7]

Akin to the standard sharing procedure, the captain gave a signal after the butchering and distribution were completed and allowed anyone to freely take the remaining meat and fat. In this case, methods (1), (4), (5), (6), (7), (9), and (11) slightly deviated from the standard procedure, as the group that successfully hunted the whale was given a share if it participated in the butchering and towing. In addition, the captain took one share more than the crew members because he owned the boat. The captain also decided how much of each part was to be served at feasts.

Case 3: J.L. (interviewed on March 7, 2010)

J.L. is a captain in his 60 s with significant whaling experience. He succeeded to his father's position as a captain. The whaling group headed by J.L. caught a whale on May 23, 2009. He distributed the catch in the following manner.

(1) *Tavsi*: J.L. cooked and served all of it at the captain's feast.

(2) *Uati*: Same as the standard sharing procedure.

(3) *Itiġruk*: Same as the standard sharing procedure.

(4) *Aqikkaak*: Same as the standard sharing procedure.

(5) *Umiat Ningingat*: Generally the same as the standard procedure, except that his whaling group, who participated in the butchering, also took a share.

(6) *Suqqaich*: The captain took half of the baleen and distributed it to his crew members or sold it. The remaining half was equally divided among the whaling groups who participated in the towing.

(7) *Sakiq*: Same as the standard sharing procedure. J.L. served the catch at the feast.

(8) *Taliġuq*: Same as the standard sharing procedure.

(9) *Utchik*: Same as the standard sharing procedure.

(10) *Uumman* (heart), *Ingaluaq* (small intestine), and *Taqtu* (kidneys): Same as the standard sharing procedure.

Akin to the standard sharing procedure, the captain gave a signal after the butchering and distribution were completed and allowed anyone to freely take the remaining meat and fat. In this case, methods (1), (5), (6), and (7) differed from the standard procedure.

The three examples presented above indicate that the captains' distribution methods varied slightly. Their sharing methods reflect their discretion, which are not necessarily consistent with the standard procedure. While captains are able to make decisions about the distribution to a certain extent, the general framework of distribution is specified by certain rules, which recommend that more food should be provided to the entire village and other whaling groups than to the captains and their crews. In the sharing practice based on butchering and towing whales, the group that hunted a whale tends to take a share if it takes part in butchering and towing. There is a case in which the captain owned the boat used for whaling and took two shares, which included one share for his contribution as a hunter (crew member) and another for the use of his boat. When the crew includes many members of the captain's direct family, the total share of the captain's household becomes large. A whaling group with members who are mostly not mutually related tends to distribute the rewards more equally among its members. Young captains in particular have demonstrated a tendency to share the catch more equally within their groups.

When the catch is distributed at the butchering site, the portions that are shared vary depending on the size and other aspects of the catch. The discretion of the captain might therefore play a greater part in such distribution. However, in many cases, members of the Barrow Whaling Captains' Wives Association supervise the butchering to prevent unfair distribution and effectively guide the sharing of meat and other parts at the buchering sites.

3.3.5 Diachronic Changes in Sharing: A Comparison Between Two Generations

Craig George conducted participant observations of bowhead whale hunts in Utqiaġvik from 1978 to 1980, when H.B.—the father of the captain interviewed in Case 2—was working as a whaling captain.

The following section presents this case (Fig. 3.6) and compares it to Case 2.

Case 4. H.B. (1978–1980)

According to Craig George's report, a whale was distributed as reported below after being butchered (George 1981).

Figure 3.6 names of the parts of a bowhead whale that are divided (prepared based on George 1981: 791)

(1) *Tavsi* (corresponding to No. 17 in Fig. 3.6): This was cooked together with the whale's internal organs and was served to the villagers during the captain's feast (George 1981: 795).

(2) *Uati* (corresponding to No. 5 and No. 6 in Fig. 3.6): No. 5 was served to the villagers at the Nalukataq festival. No. 6 was served to villagers at the Nalukataq,Thanksgiving, and Christmas celebrations (George 1981: 792).

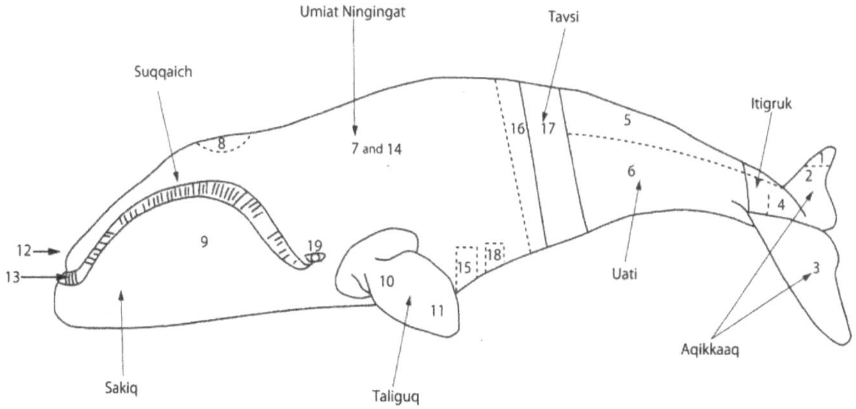

Fig. 3.2 Parts of bowhead whale (George 1981: 791)

(3) *Itiġruk* (corresponding to No. 4 in Fig. 3.6): It was served to villagers at Nalukataq (George 1981: 792).

(4) *Aqikkaak* (corresponding to No. 1, No. 2, and No. 3 in Fig. 3.6): No. 1 is a delicious part and was served to villagers at the Thanksgiving celebration in November. No. 2 was served to the villagers during the Christmas feast. No. 3 was served to the villagers at the Nalukataq festival (George 1981: 790, 792).

(5) *Umiat Ningingat* (corresponding to No. 7 in Fig. 3.6): It was divided among whaling groups that engaged in the butchering. The first whale captured was distributed equally to all registered whaling groups. All subsequent whales were apportioned only to whaling groups that engaged in the butchering (George 1981: 792).

(6) *Suqqaich* (corresponding to No. 13 in Fig. 3.6): Half of its baleen was taken by the captain of the group that successfully hunted the whale. The other half was shared equally among the other whaling groups that took part in the towing if the whale was towed by one or more boats (George 1981: 793).

(7) *Sakiq* (corresponding to No. 9 in Fig. 3.6): Half of the section from the mouth until the jaw was taken by the captain of the group that successfully hunted the whale. The other half was divided equally among whaling groups that assisted in the butchering (George 1981: 793).

(8) *Taliġuq* (corresponding to No. 10 and No. 11 in Fig. 3.6): One pectoral fin was given to the sucessful harpooner, wheras the other was given to all whaling groups that worked at the butchering site (George 1981: 793).

(9) *Utchik* (corresponding to No. 12 in Fig. 3.6): The tongue was shared among all the whaling groups that engaged in the butchering.

(10) *Uumman* (heart), *Ingaluaq* (intestine), and *Taqtu* (kidneys) (corresponding to No. 14 in Fig. 3.6): Half of it was served at the captain's feast. The remainder was stored in the captain's underground ice cellar and was served to the villagers at Nalukataq.

(11) Distribution of the other parts

No.8 in Figure 3.6: The part around the blowhole was served to villagers at the Christmas celebration (George 1981: 793).

No.15 in Fig. 3.6: *Maktak* cut into 50 cm × 50 cm × 40 cm pieces was cooked and served to the people who worked at the butchering site.

No.16 in Fig. 3.6: *Maktak* of this part was given to other whaling groups who shot shoulder guns at the whale. Alternatively, bullets with bombs for the shoulder guns were given instead of *maktak* if requested by the whalers.

No.18 in Fig. 3.6: This part (cubic *maktak* with 30-cm sides and meat of the same size) was given to the owner and operator of the pulley device used to pull the whale up onto the ice (George 1981: 794).

No.19 in Fig. 3.6: The part that includes the eardrums was taken by the captain whose group successfully hunted the whale (George 1981: 794).

Once the captain gave a signal after the butchering and distribution were completed as described above, anyone (including women) was allowed to freely take the remaining meat and other parts (George 1981: 794).

A comparison between this case and Case 2 reveals that the post-butchering distribution has been simplified, despite certain fundamental consistencies that can be observed in the distribution and use of parts. Furthermore, the use of parts such as the *Tavsi* and *Uati* has changed considerably. Although the hunters believe that their knowledge and techniques have been passed down from generation to generation, it must be emphasized that the practices have been altered over time. Moreover, the scope of a captain's discretion and the sharing procedure have been seemingly more standardized within a village.

This section explained that the sharing procedure in Utqiaġvik is determined by certain rules and that this procedure has transformed to some degree, both synchronically and diachronically. It should be emphasized at this stage that the meat and *maktak* of whales are shared among whalers who participate in the hunt and are distributed by the whalers to the entire village.

3.3.6 Secondary and Tertiary Sharing of Bowhead Whales

Following the primary sharing that takes place immediately after the butchering, whale meat and *maktak* are distributed further to other Iñupiat people within and outside Utqiaġvik through feasts, daily sharing and exchange, etc. The following section describes the secondary and tertiary sharing of whale meat and products.

3.3.6.1 Feast at the captain's Home After Successful Whaling

On the day following the completion of the hunting and butchering of a whale, a feast is held at the house of the captain whose crew caught the whale. This feast, in which whale dishes and other food are served to villagers, is called *nigipkai*. The parts of a whale that are cooked and served at this event include half of the meat and *maktak* of the belt-like part called *Tavsi* and approximately half of the whale's tongue, heart, kidneys, and small intestine.

The flag of the whaling group that caught the whale is placed on the roof of the captain's house. The crews and their wives gather at the captain's house and prepare for the feast. They cut the meat, *maktak*, tongue, heart, kidneys, and small intestine into appropriate sizes and boil them. The pectoral fins are cut into small pieces, but they are left uncooked and their fat is preserved.

These are divided into portions for the feast and portions for gifts. The meat and internal organs for gifts are divided into amounts commensurate to one meal for one person and are packed in plastic bags. They also prepare bread shaped like muffins, fried bread (donuts), cooked fruits, coffee, and tea.

When the preparation is complete, the captain uses a radio to pray and thank the Christian God for the successful hunt. He subsequently announces the start of the celebration and calls the villagers to take the whale meat and other food. When the villagers come to visit his house, the captain and the wives and daughters of his whaling group hand the food packed in plastic bags to the villagers. The villagers then take this food home and eat it. The elders and the captain's relatives enjoy the whale meat and *maktak* in the living room of the captain's house and then take whale meat and other food home as souvenirs.

The captain and others deliver the packed food to several elders and widows who are unable to visit the captain's house.

When all the food prepared for guests has been given out or consumed, the captain announces the end of the feast over the radio and takes down his whaling group's flag from the roof.

Through this feast, single-meal packs of whale dishes are provided to many households in the village. Slightly more whale meat and other food are provided to the households of widows and elders.

3.3.6.2 Sharing at Festivals

After the spring whaling, festivals called Apugauti and Nalukataq are held. During these festivals, whale dishes are served to villagers. Apugauti is held independently by each group that successfully hunted whales in the year, on the day they lift up an *umiaq* (large boat made of seal skin) to the land in the end of whaling season. It is held approximately from mid-May to early June. After the Apugauti, the Nalukataq festival is held several times by a single or multiple whaling groups working in cooperation from mid-June to late July (Kishigami 2011b).

While some captains do not hold Apugauti, it is held more commonly for the same number of days as number of whales caught in the spring. The entrée of the Apugauti celebration is a dish called *mikigaq*, which is made by fermenting a mixture of whale meat, blubber, blood, etc. and is served with duck soup/goose soup, bread, fruits, and coffee/tea. For one particular Apugauti, enough *mikigaq* to fill 14 plastic buckets (266 L), 20 pots each of duck soup and goose soup, 2000 pieces each of bread and Eskimo donuts, 10 buckets of simmered fruits (190 L), approximately 53 L of tea

Photo 3.4 Community feast at Nalukataq. Barrow (Utqiaġvik), Alaska, USA. June, 2012. By N. Kishigami

and approximately 38 L of coffee were prepared. This was a rather large celebration. The amount of whale dishes served at an Apugauti feast depends on the decision of the captain. A small feast might have about 100 guests. A large feast might be attended by more than 400 villagers. Each guest is given enough food for one or two meals (Kishigami 2011b) (Photo 3.7).

In Utqiaġvik, Nalukataq celebrations are held several times in a year. The Nalukataq festival is organized by captains who have caught whales during the spring of the same year. In a year in which numerous whales have been landed, Nalukataq is held four or five times per summer season. In Utqiaġvik, each Nalukataq celebration lasts one day. On June 30, 2008, two captains jointly held a Nalukataq celebration. At noon, duck soup, goose soup, bread, coffee, and tea were served. *Mikigaq* was served at three in the afternoon. Whale meat and *maktak*, cakes, fruits, tea, and coffee were served at six in the evening. The captains set aside approximately a third of the *Uati*, *Itigaruk*, and caudal fins and half of the heart, kidneys, and small intestine of the hunted whale for this feast. This gathering is enjoyed by more than 600 guests each time (Stephen R. Braund and Associates 2018: 23), but each guest receives only one meal each at noon, three in afternoon, and six in the evening, respectively. If a large amount of whale meat and fat is available, it is divided and distributed to guests as souvenirs. The blanket toss, which is a play activity similar to trampolining,[8] and the Iñupiat's traditional dance are performed after the Nalukataq feast.

During Thanksgiving and Christmas, feasts are held at multiple churches in the village. Approximately a third of the *Uati*, *Itigaruk*, and caudal fins of the whales hunted in the year are served at each feast. More than five denominations of Christian churches exist in the village. Captains serve the dishes to three to five churches, which include the Presbyterian Church, Cornerstone Church, and Assemblies of God Church. If a large amount of frozen whale meat and *maktak* is available on Thanksgiving or Christmas, it is divided equally among those who attend the feast. In addition, during the Messenger Feast (*Kivġiq*) that is held once in two years or more,[9] captains and hunters contribute whale dishes and invite people from other villages. More than 20 such public feasts, where whale meat, *maktak*, internal organs, and *mikigaq* are provided to and consumed by the villagers, are held each year.

3.3.6.3 Sharing of Whale Meat and Blubber Within a Village

Once the whale meat and *maktak* are distributed among the captain and hunters in a whaling group, they are shared with other people. The captain and hunters provide a part of their share to their parents, siblings, uncles, aunts, and other members of their extended families, parents in-laws, and friends, in addition to sharing the whale meat and *maktak* through meals.[10]

The captain and hunters share a part of their whale meat and *maktak* with elders, widows, sick people, and households in need, regardless of whether they are related to them genealogicaly or by marriage.[11] They occasionally give whale meat and *maktak* as a gift to a family that has recently lost a member. They sometimes give whale baleen to the relatives of craftspeople.

3.3.6.4 Sharing of Whale Meat and *Maktak* Outside a Village

Almost all whaling captains and hunters in Utqiaġvik have either siblings, cousins, children, nieces, nephews, or friends who live in other villages or states. They communicate with each other via telephone, e-mail, and social media such as Facebook, WhatsApp, etc. They send whale meat and *maktak*, caribou meat, and other local products by airmail or entrust a person deliver these items to relatives and friends who live not only in neighboring villages such as Kaktovik and Atqasuk, but in more distant towns such as Point Hope, Kotzwbue, Anchorage, Fairbanks, and Seattle as well. They occasionally send whale meat and *maktak* to a family member or relative hospitalized in Anchorage. They even donate whale meat and *maktak* to prisons in Fairbanks. I have been able to confirm that gifts were sent to Nuiqsut, Atqasuk, Point Lay, Point Hope, Selawik, Fairbanks, Anchorage, Nome, Anaktuvuk Pass, Kotzebue, Noorvik, and the states of Montana, Washington, Arizona, California, and Hawaii. These locations reflect that the Iñupiat people migrate to many parts of the United States for jobs, education, marriage, and other reasons.

According to the 2010 Census, 27% of the total Iñupiat population had migrated to urban areas in Alaska and 23% had migrated outside the state. This dispersal explains why whale meat and *maktak* is sent from the Iñupiat's whaling villages to those living in various locations outside the villages (Kofinas et al. 2016; Braund and

Associates 2018). For example, approximately 25–30% of Nuiqsut's whale products are estimated to be sent to places outside the village and approximately 30–40% of Point Hope's whale products are presumably transported outward (Stephen R. Braund and Associates 2018: 54).

Families who live outside their home villages often send smelt, whitefish, and beluga *maktak* to their relatives back home. For instance, I observed that a hunter living in Kaktovik sent a bucket full of *maktak* to his cousin in Utqiaġvik through a third person when he caught a 12 m long bowhead whale.

Some of them promise an exchange of local products to the people whom they meet at various business meetings of indigenous organizations. For instance, an Iñupiat man exchanged dried salmon for approximately 20 kg of whale meat and *maktak* with a person from another indigenous group, whom he met at a meeting for Alaskan indigenous people.

The following cases of product exchange have been confirmed through my research (Table 3.1).

As outlined above, sharing and exchange with people outside the village are actively conducted. Many reports have described cases in which people from a non-whaling community provide a whaling community with goods and cash and receive a share of whale meat and *maktak* upon successful whaling (Stephen R. Braund and Associates 2018; Kofinas et al. 2016).

Table 3.1 Exchange of food and goods between Utqiaġvik and other villages/cities

Persons outside Utqiaġvik	Food/goods to Utqiaġvik	Food/goods from Utqiaġvik
A non-Iñupiat friend in Hooper Bay, Alaska	Moose meat	*Maktak*
A non-Iñupiat friend on Kodiak Island	Salmon and halibut	*Maktak*
A non-Iñupiat friend in inland Alaska	Caribou fat	Whale meat, *maktak*, whale oil
An Iñupiat friend in Wainwright	Arctic rainbow smelt, beluga whale meat	*maktak*, caribou meat, white fish, bearded seal meat, walrus meat
A non-Iñupiat friend on Yukon River, inland Alaska	Salmon, moose stripes	Seal oil, *maktak*
A non-Iñupiat friend in Bethel	Salmon, berries	*Maktak*, seal oil
An Iñupiat friend in Atqasuq	White fish with roe	*Maktak*
Iñupiat friends in Anchorage and Fairbanks	Tea, coffee, flour, gloves, rope for a float	Whale meat, *maktak*

3.3.6.5 Sale and exchange of Whale Meat and *Maktak*

Whale meat and other edible parts possess symbolic value for the Iñupiat and are not sold for cash. They are supplied to villagers through rule-based sharing and voluntary sharing (Bodenhorn 2000: 38). The villagers share whale meat and other parts with their families, relatives, and friends living outside the village and exchange them with products from other regions. As whaling captains customarily do not sell whale meat and other parts to other Iñupiat people, they cannot make money from whaling. Thus, they carry out whale hunts using their own funds.

A whale product that is commercially traded is baleen. Approximately 30 cm of the baleen plate is sold for US$35. However, captains and hunters often provide whale baleen and bones free of charge to their families, relatives, and friends who are craftspeople. The village's craftspeople sell items made of whale baleen and bones to tourists.

3.3.6.6 Sharing Within a Whaling Group

Hunters in Utqiaġvik prefer to hunt whales that are approximately 10 m long. The following section uses existing data to examine the proportion of each part of a whale in terms of weight.

George et al. reported the weight of each part (excluding bones, etc.) of a male whale that had a total length of 11 m and a total weight of 14,797 kg, which is illustrated in Table 3.2 (George et al. 1988). This 11 m long bowhead whale represented the average size of whales hunted by the Iñupiat.

These proportions of the weights of its parts are regarded as generally comparable to those of bowhead whales of other sizes (George et al. 1988). The ratios

Table 3.2 Weight and percentage of each part of an 11-m-long bowhead whale

Name of Part	Weight	Percentage of weight (%)
Tongue	893 kg	(6)
Maktak (skin parts with fat)	6601 kg	(44)
Meat	2428 kg	(16)
Baleen	595.5 kg	(4)
Caudal fins	217.7 kg	(1.5)
Pectoral fins	349.2 kg	(2.4)
Kidneys (both)	97.9 kg	(0.7)
Heart	95.2 kg	(0.8)
Small intestine	223.8 kg	(1.5)

Source George, Philo, Carroll, and Albert (1988)
* Between a quarter and a fifth of the total *maktak* is edible
* The weights of the whale's bones, blood, and lungs are excluded from this table

of the weights of the *Umiat Ningingat*, *Tavsi*, and *Uati*, at the time of butchering were broadly 60%, 10%, and 30%, respectively (Brower and Hepa 1998: 38). The following table estimates the weights of the meat and *maktak* based on these ratios. The *maktak* of the edible parts was assumed to be 20%.

A female whale that had a length of 8.4 m and an estimated weight of 11,300 kg was captured in a whaling on May 4, 2010 (6 in Table 3.3). Calculations based on Table 3.2

Table 3.3 Harvest day, Length, sex of bowhead whales landed in Utqiaġvik in 2010

Number	Day/month of harvest in whale hunts in 2010	Length of whales landed (meters)	Sex of landed whale
Spring whale hunt			
1	May 1	10.9	Female
2	May 3	8.3	Female
3	May 4	8.0	Female
4	May 4	8.7	Male
5	May 4	8.7	Male
6	May 4	8.4	Female
7	May 5	8.4	Male
8	May 5	7.3	Male
9	May 5	8.7	unidentified
10	May 6	10.7	Female
11	May 7	7.5	Male
12	May 9	9.8	Female
13	May 12	13.1	Female
14	May 15	8.3	Female
1 ~ 14	Average length of the whales landed in spring	9.1	Male:5, female:8, unidentified: 1
Fall whale hunt			
15	October 7	12.3	Female
16	October 7	7.8	Male
17	October 7	10.8	Male
18	October 8	9.0	Female
19	October 8	10.9	Female
20	October 9	7.7	Male
21	October 9	11.2	Male
22	October 11	7.1	Female
15~22	Average length of the whales landed in spring	9.6	Male: 4, female: 4
1~22	Average length of the whales landed in 2010	9.3	Male: 9, female: 12, unidentified: 1

Source Suydam et al. (2011:6)

indicate that approximately the 1808 kg of meat collected from it comprised 16% of its total weight and that the edible part of the *maktak*, which weighed approximately 994 kg, comprised 8.8% of its total weight. The *Umiat Ningingat* that were distributed to whaling groups and individuals who took part in the butchering consisted of approximately 1085 kg of meat and 596 kg of *maktak*.

This whale was butchered by a total of 15 whaling groups—each of which sent two or three helpers—and four independent individuals from the community. The four individuals were considered to be one whaling group. Thus, the parts to be shared by the butchers were divided into 16 shares. Each whaling group that participated in the butchering presumably received approximately 68 kg of meat and 37 kg of *maktak*. The captains of these whaling groups would redistribute the meat and *maktak* to the members of their groups.

The *Tavsi*, which was consumed at the captain's feast and was distributed to the whaling group that captured the whale, consisted of approximately 90 kg of meat and 50 kg of *maktak*. In addition to the meat and *maktak*, approximately 170 kg of tongue, 40 kg of kidneys, 34 kg of heart, and 85 kg of small intestine were cooked and served to villagers at the captain's feast.

The *Uati*, which was provided to the entire community during the Nalukataq festival, Thanksgiving, and Christmas, consisted of approximately 542 kg of meat and 298 kg of *maktak*. The captain decides the proportions of meat and *maktak* that are served for the feasts at the festivals. Assuming that the meat and *maktak* alloted to be served the Nalukataq, Thanksgiving, and Christmas feasts were 50%, 25%, and 25%, respectively, approximately 271 kg of meat and 149 kg of *maktak* would be served at the Nalukataq festival. The Thanksgiving and Christmas feasts would resultantly be alloted approximately 136 kg of meat and 75 kg of *maktak* each. Additionally, approximately 170 kg of tongue, 40 kg of kidneys, 34 kg of heart, and 85 kg of small intestine would be provided during Nalukataq.

A total of 22 whales were caught in 2010, as shown in Table 3.3. Based on their length, the weight of the total catch would be 273,556 kg.[12] The edible parts would consist of 43,769 kg of meat and 24,073 kg of *maktak*. From these, more than 13,130 kg of whale meat and 7222 kg of *maktak* would be consumed or shared at village feasts.

Most parts of the hunted whales are provided for feasts held for the entire community and for other whaling groups and individuals who participate in the butchering. The group that catches the whale rceives half of the *Tavsi* and one share of the *Umiat Ningingat*. These are divided within the group as per captain's discretion. In the case of May 4, 2010 (6 in Table 3.3), the meat and *maktak* were shared equally among eight persons, which included five crew members (one captain and four subordinates), two fuel providers, and one lender of darting guns. In this case, there was no difference between the share taken by the captain and that given to the crew members.

Each person's share was approximately 19.8 kg of meat and 10.9 kg of *maktak*. The harpooner was given approximately 158 kg of a pectoral fin. Thus, it was revealed that the amount obtained freely by the members of the group that had captured the whale was not very large.

However, when other groups caught a whale, the group would receive appropriate amounts of meat and *maktak* from the *Umiat Ningingat* part if it sent helpers to assist the butchering process. The following estimates the total amounts of meat and *maktak* that the group in this case would receive if it caught one whale and participated in butchering on 10 occasions. As shown in Table 3.3, the average length of whales caught in 2010 was 9.3 m. The total weight per whale is therefore estimated to be 12,510 kg. Assuming that 16 groups participated in the butchering, each group would receive approximately 75 kg of meat and approximately 41.3 kg of *maktak* each time. Thus, the group would gain a total of 750 kg of meat and 413 kg of *maktak* if it assisted in the butchering of whales caught by other groups 10 times during the year. Therefore, each person would receive approximately 93.8 kg of meat and 51.6 kg of *maktak* if this is shared among the above eight people.

In the case of the aforementioned whaling group, if the meat and *maktak* were shared equally among the captain and crew members, then each person would obtain a total of approximately 113.6 kg of meat and 62.5 kg of *maktak* every year. This meat and *maktak* can be consumed and shared freely by each of their households. It should be noted that if a group has multiple members (hunters) who belong to the same household or extended family as the whaling captain, then the share of their household or extended family will be larger than the share that other members receive. The captain's share will also be larger if they gets an additional share for the boat.

This section explained that due to the rule-based distribution, approximately 95% of whale meat is provided to villagers and whaling groups other than the group that catches the whale. In other words, only slightly more than 5% of all whale meat and *maktak* is exclusively available to the group that catches the whale.

3.3.7 Sharing and Giving

Whale meat, *maktak*, and other edible parts (except baleen) are not sold. Instead, they are shared or exchanged with other products. The primary distribution of whale products after the butchering and the provision of these products during the captain's feast and festivals such as Apugauti,[13] Nalukataq, Thanksgiving, Christmas, and Messenger Feast are conducted in accordance with certain rules. Other types of distribution, such as the secondary sharing, giving, and exchange, are conducted voluntarily.

As described earlier, the methods of sharing after the butchering vary across groups, whether within Utqiaġvik or in other villages. This is because the diachronic and synchronic changes in the sharing practices in Utqiaġvik are affected by the discretion of the captains. Worl and Smythe (1986: 300) pointed out that, as of 1986, the frequency of daily food sharing in Utqiaġvik had been decreasing and the good-intentioned distribution of food without any expectation of reciprocity was gradually replaced with barter-like exchange. Despite these changes, the rule-based primary distribution of whale meat and *maktak* and food sharing within extended families continue (Bodenhorn 2000). The sharing of whale dishes through various feasts is also practiced in accordance with specific rules.

The next section will present the characteristics, reasons, and functions of the sharing and giving of whale products by the Iñupiat.

3.3.7.1 Characteristics of Sharing a Whale

Many Alaskan Iñupiat engage in whaling according to their tradition and within the framework of the Aboriginal Subsistence Whaling established by the IWC, and share and consume the products they obtain from whaling. They do not sell whale parts (except baleen) for money, but share them free of charge or exchange them with other products.

Two characteristics are observable in the sharing of whales by the Iñupiat. One is that there are two types of sharing: one is based on clear rules and the other is done voluntarily. The former consists of the primary sharing that is conducted after the butchering and the communal feasts. While the discretion of the captains influences the sharing to some extent, there are clear rules regarding how a whale is to be shared and which parts are to be provided for community feasts. Personal sharing and exchange by captains and their hunters depend on their intentions.

Due to this sharing system, a crew that has successfully hunted a whale is unable to obtain a large amount of meat and *maktak* from their own catch. Thus, the second characteristic is that most of the catch is shared during community feasts or is provided to other whaling groups that assist in the towing and butchering of the whale. Moreover, whale meat and *maktak* are shared among Iñupiat people living within and outside the village through personal sharing, giving, and exchange.

3.3.7.2 Reasons for Rule-Based Sharing

In this section, I will discuss the reasons for the rule-based sharing of whale products. Studies related to this research include that of Bodenhorn (2000), who studied the food sharing practices of the Iñupiat, and Dahl (2000), who studied the food sharing practices of the Greenlandic Inuit.

Bodenhorn (2000: 28–29) argued that shares and sharing should be distinguished when assessing the sharing practices related to whale products. The difference is that a share is given in return for performing specific work or fulfilling responsibilities. However, with respect to sharing, people can "decide for themselves how much they are going to give, to whom and in what context" (Bodenhorn 2000: 39). She claims that the former is given upon providing a means of production or participating in hunting, while the latter is profoundly related to the worldview of the Iñupiat. The Iñupiat believe that "the more one gives away, the more the animals are likely to give themselves up in return" (Bodenhorn 2000: 44) and "animals come to those who share" (Bodenhorn 2000: 47). Thus, they consider the practice of food sharing to be significant. Further, Bodenhorn (2000: 48) asserted that food sharing is a means of obtaining country food such as whale meat, which is not available for purchase with money in the current socioeconomic context, and is one of the modes of adapting to the contemporary conditions.

Dahl conducted research in Saqqaq in the northwestern region of Greenland and argued that sharing should be distinguished from the giving of a meat gift. The village in this region has three types of sharing: *ningerpoq*, which is traditional sharing of game; *agguarpoq*, which is sharing of beluga whales by boat crews that has been institutionalized in recent years; and *pajuppoq*, which is a meat gift. Dahl states that the first two are a part of a production relation. The third belongs to a social relation or social exchange system. The first two are sharing based on rules derived from the rights of hunters to specific parts, meat, and *maktak* of beluga whales, whereas the third is voluntary sharing based on social ethics that do not demand reciprocity. In other words, Dahl (2000: 175–178) argued that there are two types of sharing: one is specified by rules and is based on a production relation, whereas the other is voluntary and without obligations. An interesting aspect of the claims of Bodenhorn and Dahl is that while they use different descriptive concepts, the meaning of the term "share" used by Bodenhorn and that of "sharing" used by Dahl are nearly the same. Even in different regions, the sharing of bowhead whales and beluga whales is regulated by rules when it concerns a production relation or co-production.

In the whaling done by the Iñupiat, while the killing of a whale can be executed by one whaling group, it is unlikely that the group alone can complete the entire series of whaling activities, which includes towing the whale to the butchering site, placing it on ice or land, butchering it, and storing it. In nearly all cases of whaling, the activities meant to be completed after the capture of a whale require the participation of two or more whaling groups. Thus, the rules for sharing likely exist in order to give the providers of labor and tools appropriate rewards for their contribution. In addition, these rules prevent the monopolistic ownership of captured whales. Instead, they facilitate the sharing of a catch with other whaling groups and villagers. The rules thus function as a social mechanism that ensures that whaling activity turns out to be widely beneficial for the entire Iñupiat community in Utqiaġvik. I therefore argue that the primary reason for establishing the sharing rules is to create a mechanism that will contribute to the welfare of the entire community.

3.3.7.3 Reasons for and Functions of the Continuance of Whale Sharing

Why do captains and their crew members continue to share whale products despite the lack of opportunities to gain significant personal profits from successful whaling?[14] As mentioned in the first chapter, several studies have been conducted to elucidate the origin and existence of food sharing. Although this study will not attempt to verify all the existing hypotheses, I believe that it is appropriate to examine the social functions and effects of food sharing for the Iñupiat in particular, in order to explain their reasons for continuing the practice. In addition, I will also devote attention to differences between rule-based sharing and voluntary sharing practices.

Currently, Utqiaġvik is home to numerous European Americans and immigrant workers from Thailand, South Korea, and the Philippines, in addition to the Iñupiat. However, the sharing of whale products such as whale meat and *maktak* is generally limited to the Iñupiat community.

First, in rule-based sharing, whales (food with high cultural value) are provided to the entire village through feasts. Whaling captains and hunters emphasize that they hunt whales to provide food to the entire community, which is regarded as an important objective of their whaling. As pointed out earlier, nearly 95% of edible parts are provided to other whaling groups that helped tow and butcher the whales (approximately 60%) and are served at feasts held for the Iñupiat villagers, such as during the Apugauti and Nalukataq festivals (approximately 35%). Successful whaling brings meat and *maktak* not only to other whaling groups, but also to the entire Iñupiat community through various community feasts. The feasts serve as an opportunity for the Iñupiat to celebrate and remind themselves of their identity by comunally consuming whale dishes that possess significant cultural value.

Second, the continued practice of rule-based sharing and voluntary sharing helps reduce disparities in the consumption of whale products across households, extended families, and whaling groups. Villagers that seek whale meat are able to obtain and consume it by attending feasts at the houses of successful whaling captains or during Apugauti, Nalukataq, Thanksgiving, and Christmas. Additionally, they also receive whale meat and *maktak* from whalers' families, relatives, and friends. Although there are differences in the amounts of whale meat and *maktak* owned by captains' households, crew members' households, and households without whalers, rule-based sharing and voluntary sharing practices reduce disparities and facilitate a somewhat egalitarian distribution of whale meat and *maktak* within the village.

Third, sharing through various community feasts and sharing within each whaling group are related to reproduction of identity and social relationships at the level of the entire village, at the level of whaling groups, and at the level of extended families. Voluntary sharing confirms and maintains family and kin relationships through the distribution of whale products along specific social relationships from certain individuals (households) that have whale meat and *maktak* to other people (households).

Fourth, voluntary sharing provides social security by reducing the food shortages of the socially vulnerable, such as widows, the elderly, and people in need, which contributes to the welfare of the entire community. As already explained, captains and hunters of whaling groups are able to obtain approximately 113.6 kg of whale meat and approximately 62.5 kg of *maktak* per person through a year of whaling activities. The whale meat and *maktak* that they obtain are consumed not only in their own households, but are also distributed to their families, relatives, friends, widows, elderly people, and those in need of food. It should be emphasized that whaling captains and hunters, in particular, ensure that large amounts of whale meat and *maktak* are given to the socially disadvantaged.

Fifth, the rule-based sharing and voluntary sharing are profoundly related to the perception of the Iñupiat with respect to their relationship with whales, which is reproduced through the sharing practices. Bodenhorn (1990, 2000, 2005) pointed out that the Iñupiat believe that they have a special relationship with whales and other animals, which forms the background of voluntary sharing. I believe that rule-based sharing is also deeply related to this worldview. This rule-based sharing is a system that enables the entire community to enjoy the catch, rather than simply the successful captain and their crew (hunters). The successful whaling group provides most of the catch to other people and only retains some of it. This practice is truly based on the relationship between whales and the Iñupiat. Whales are special beings for the Iñupiat. The Iñupiat believe that whales have the ability to detect what is occurring in human society and offer their lives to the captains who treat whales and other people in an appropriate manner. They believe that whales are attracted those captains/captains' wives who display a proper attitude towards whales. In particular, the captains and their wives must treat both whales and fellow humans desirably. The Iñupiat believe that whales will return to be captured by the same whaling captain if they continue to refrain from acting in ways disliked by whales and share their catch with others. They engage in whaling while simultaneously disciplining themselves through worldview (Bodenhorn 1990; Turner 1990; Victor 1987).

Although certain general principles are maintained, the relationship between the Iñupiat and whales has been changing because of the influence of Christianity. As the Iñupiat believe that God sends whales to them, they now express their gratitude to God instead upon successfully completing a hunt (Kishigami 2009b). Their view of whales has been changing, but whaling remains closely connected to the Iñupiat's worldviews. Their whaling and sharing of whale products still presumably reflect their worldviews.[15]

Sixth, rule-based sharing as well as voluntary sharing increase the reputation of the captain and hunters who capture a whale. The Iñupiat in Utqiaġvik have a sharing system in which a whaling group that succeeds in its hunt provides most of its catch to other whaling groups and villagers. Captains and thier crews (hunters) who always succeed in whaling and hold feasts are considered "true Iñupiat" and are respected by the villagers. Although they are not able to earn monetary gains, they attain social reputation through their whaling and sharing. A captain and his/her crew are thus able to gain cultural satisfaction through successful whaling and sharing of a catch. The captains and captains' spouses who are deemed to be significant contributors to the community tend to be elected as mayors of a borough, members of a borough assembly, village mayors, and/or directors of indigenous organizations. They conceivably have political influence over other villagers.

For the Iñupiat, whaling is a kind of sacred activity. I presume that the process of whaling and the subsequent sharing of whale products promotes a festive mood and creates great excitement—similar to Javanese cockfights (Geertz 1973), the Gion Festival in Kyoto, or the Sanja Festival in Toyo. Kruse (1991), who studied Iñupiat subsistence activities and wage labor, reported that they were more attracted to subsistence activities than wage labor. He claims that instead of monetary profit, subsistence activities can bring about process benefits for the Iñupiat, such as social interaction, a sense of challenge, a sense of achievement, and time away from daily routines (Kruse 1991: 324–325). Although Kruse's research dates back more than 30 years, I believe that this idea might still apply today to a considerable extent. By participating in whaling and the sharing of a catch, the contemporary Iñupiat are seemingly gaining a sense of cultural achievement, satisfaction, and other psychological rewards that they do not gain from wage labor.

As discussed until now, the Iñupiat's whaling and subsequent sharing of a catch demonstrate multiple functions and effects, which appear to be the reasons for the persistence of these sharing practices. I believe that the key reasons for the preservation of these sharing practices are the contribution of this sharing to the well-being of the entire community and the improvement of whalers' social prestige and cultural satisfaction. It should also be emphasized that whaling and the sharing of a catch contribute toward making their life meaningful.

3.4 Sharing of Food Other Than Whale Meat

In this section, I will describe the sharing practices concerning food other than whale meat in Utqiaġvik.

3.4.1 Sharing of Game

The sharing practices vary depending on the game caught. Animals and fish caught by a person become their property. A part of the catch is shared with people who participated or assisted in the hunting or fishing process. For instance, if fish are caught using a fishnet, a part of the catch is shared with anyone who helped set the fishnet, monitor the site, remove fish from the net, or performed other such activities. Waterfowl become the property of the person who catches them, but is shared with any person who helped pick up and collect the birds.

A part of the game is provided to those who participated in the hunting, provided hunting tools, or otherwise contributed to the means of production. If a ringed seal, bearded seal, or walrus is caught, then the owner of the boat used for the hunting receives an additional share and the hunter who caught the animal receives extra skin, tusk, etc. The sharing is thus unequal. If a beluga whale is caught by a group of hunters, its meat and fat are distributed within the group. A hunter who has killed a polar bear obtains its skin, but its meat is shared with others (Bodenhorn 2000: 32).

Two hunters who go hunting together and catch an animal share the catch between them. Each hunter subsequently gives a part of his share to people who lent them a rifle and outboard motor and/or provided gasoline (Bodenhorn 2000: 32).

The hunters distribute meat and fat from their own share to other families, relatives, and villagers. They occasionally also provide some of it to their families, relatives, and friends who live outside the village. Although hunters are expected to share meat and other products with others, they should decide how much to give and with whom (Bodenhorn 2000: 39).

3.4.2 Sharing at the Time of Butchering

When butchering a caribou or walrus at a hunting site, meat and other parts are shared with the hunters who helped with the butchering.

When several women butcher a caribou at a camping site or in a village, the caribou meat is shared equally among these women. The hunter who has captured the caribou acquires its fur, which used to be a trade good and a source of wealth in the old days (Bodenhorn 2000: 32).

If a hunter brings home several bearded seals or ringed seals, his wife and/or mother gather women and have them help with the butchering. After the seals are butchered, their meat and fat are apportioned to the women who assisted. Widows and women from households that have no hunters are often intentionally selected as helpers by the hunter's wife and/or mother (Bodenhorn 2000: 33).

3.4.3 One-Sided Sharing

When a beluga whale, polar bear, or bowhead whale is distributed to the entire village, any villager might receive a portion of the meat and take it home. Villagers sometimes go and receive meat after being told to do so by the hunter or his family. In such a case, they are not obligated to return anything to the hunter. Any villager who is in need of food might request another villager for food. In such a case too there is no obligation of reciprocity.

Hunters and meat owners provide meat to elderly people, widows, sick people, single people, and others, regardless of whether they share a familial relationship. Hunters and others at times provide meat to women's shelters and senior citizen homes in the village and prisons in Fairbanks. Such distribution is one-sided, which means that no return is expected (Bodenhorn 2000: 43).

3.4.4 Characteristics of Sharing

The meat (*niqipiaq*) acquired through hunting by the Iñupiat can be shared between hunters and other people either voluntarily or in line with the sharing rules.

A hunter provides a part of the catch to the people who played a specific role in the hunt/fishing and people who contributed to the hunt/fishing by providing tools, gasoline, and other necessary items. The voluntary sharing applies primarily to the hunter's relatives and to socially vulnerable people. In Iñupiat society, sharing networks are formed among family and kin households and food is exchanged mutually within such networks in most cases. Sharing is conducted most frequently among families and relatives, which takes a form similar to reciprocal exchange (Bodenhorn 2000: 42). Meat is supplied one-sidedly to elderly people, widows, households without hunters, and other socially vulnerable people, regardless of whether they are part of a sharing network or not.

The Iñupiat sharing practices reflect their worldview of relationships between animals and humans. Animals offer their lives to people (hunters) and become food for them. The Iñupiat believe that the same animals will be repeatedly caught by the same hunters, provided they are treated properly. The Iñupiat believe that a hunter's propensity to hunt successfully is commensurate to their generosity. Sharing helps maintain and reproduce social relationships not only between people and animals, but also among people. Sharing is a means for them to express their ethnicity. Sharing allows the Iñupiat to have access to country ('true') food (Bodenhorn 2000: 44–48).

According to their worldview, sharing is neither mutually rewarding nor demand sharing (Bodenhorn 2000: 49). The Iñupiat perceive their sharing to be a positive and virtuous experience. In other words, Iñupiat hunters gain satisfaction and joy through hunting and sharing (Bodenhorn 2000: 27).

3.5 Conclusion

This chapter has reported and discussed the Iñupiat's food sharing practices. Bodenhorn's research on the sharing of bowhead whales and the research by Dahl and Graburn on the sharing of beluga whales reveal that the primary sharing after capturing a large animal is based on rules (Bodenhorn 2000: 28, 35–36; Dahl 2000: 175–178; Graburn 1969: 68–70). The sharing of large animals is apparently based on a logic that is different from the sharing of seal meat and caribou meat among Inuit, which was described in Chap. 1 and 2 (Hovelsrud-Broda 2000; Kishigami 2000; Wenzel 2000). The cooperation hypothesis and show-off hypothesis are often used to elucidate the reasons for the rule-based sharing of large animals. However, the latter hypothesis has been disproved by research concerning evolutionary ecology (Alvard 2002; Alvard and Nolin 2002).

The Iñupiat's sharing of whales can be classified broadly into two types: rule-based sharing and voluntary sharing. The rule-based sharing practices vary across villages, within a village, and across different generations in the same village. The diachronic and synchronic differences within the same village are presumably created by the discretion of whaling captains.

Although there may be slight differences between the whaling groups, the rule-based sharing primarily functions as a mechanism of distributing food that possesses high cultural value, which is the result of cooperative labor, to the entire community. Furthermore, the two types of sharing have various functions and effects, such as contributing to the acquisition of food with high cultural value and the well-being of the community and facilitating the egalitarian distribution of food. Additionally, it reproduces worldviews, identities, and social relationships at various levels. Whaling captains and their crews (hunters) are able to gain both social reputation and cultural satisfaction through practice of sharing. I argue that the Iñupiat in Utqiaġvik continue to practice the two types of sharing because of these functions and effects.

The practice of whaling and the sharing of catch in modern Iñupiat society is a pursuit of valued resources for the community benefit, not for personal gain. The distribution of such resources throughout the entire community has become the aim of the Iñupiat hunters. It is thus likely to contribute to the well-being of the entire community, as in the case of the Canadian Inuit (Kishigami 2007; Wenzel 1991: 100).[16]

Notes

1 Whale baleen was an important product in Alaskan indigenous people's trade with Europeans and Americans from the end of the nineteenth century until the early twentieth century. It brought great economic wealth to whaling captains (Bockstoce 2009; Sheehan 1997). Presently, however, whale baleen and bones are rarely sold or bought for money.

2 Whalers in villages such as Utqiaġvik, Kaktovik, and Nuiqsut tend to hunt small whales that are approximately 10 m long, which are easy to catch and have soft and tasty meat. Whalers at Point Hope and Wainwright tend to hunt larger whales that have a length of more than 13 m, each of which brings a large amount of meat and *maktak* to the village.

3 Cash income from wage labor, dividends given by Alaska's indigenous organizations and Alaska's state government, and income from other sources are used to conduct whaling activities (Kishigami 2009a; Kishigami 2010). Some Iñupiat people sell whale baleen to local craftspeople. However, the baleen are more often provided free of charge. Whale baleen were sold to traders who came from Europe and America during commercial whaling (around 1848–1914), which can thereby be considered a part of traditional activities in a broad sense.

4 Kishigami (2009a, 2010) examined the effects of global warming, exploration for oil, the use of the Northwest Passage for transportation, and the animal and environmental protection movement on whaling activities.

5 The term "share" means a portion to be taken. If whale meat is equally distributed among 12 groups, then a twelfth of the total amount of meat represents one share.

6 Fall whaling often takes place in an area more than 10 km off the coast of the village. Thus, multiple engine-powered boats must be used to tow a captured whale to a coast on the outskirts of the village where the butchering site is located. During spring whaling, whales are captured near the edge of the sea ice, pulled up onto nearby landfast ice, and butchered. In many cases, engine-powered boats are used.

7 Eardrums are symbolic memorials and serve as a record of captured whales.

8 Blanket toss is a trampoline-like play activity that uses a sheet made by sewing together boat covers made of seal skins, which are used for an *umiaq* boat. The *umiaq* crew members and their family members hold the edges of the sheet, and each person takes a turn and jumps high on the sheet.

9 Studies on the current Messenger Feast include research carried out by Ikuta (2007).

10 Detailed studies of food distribution and exchange in daily life have been conducted by Bodenhorn (2000: 38–44; 2005).

11 Someone who personally owns whale meat and *maktak* voluntarily gives them to others in some cases. In other cases, they might give them to others upon demand by receipents.

12 The statistics from 2003 indicate that the total population of Utqiaġvik was 4,429. Of the total population, 2715 people (61.3%) were Iñupiat (Kishigami 2009a: 506–507). Bowhead whales comprised approximately 19% of all food consumed by the Iñupiat in 1977 (Donovan 1982: 81). Although details are not available, this percentage is estimated to have further declined by 2000, when the population had nearly doubled. In other words, I bleieve that whale meat and *maktak* have become more symbolic dishes rather than a part of the everyday diet of the Iñupiat. However, at many households of middle-aged and elderly people, whale meat and *maktak* are consumed on many occasions, even if quantity consumed is small.

13 Apugauti began to take its current form in the 1980s, but some captains do not observe it. Originally, Apugauti was a small feast held for the whaling group upon their return to land with their boat and camping gear. It involved serving special dishes such as *mikigaq* in order to greet an exhausted captain and crew (hunters) at a beach (Kishigami 2011b).

14 From 1848–1916, Iñupiat people were able to profit from selling whale baleen to traders and accumulated wealth. Before that period, they accumulated wealth using the indigenous trade networks across Eurasia and Alaska and within Northern North America (Bockstoce 2009; Sheehan 1997).

15 An ideology which considers the relationship between humans and animals as being mutually beneficial exists all across the arctic area that ranges from Siberia to Greenland. Fienup-Riordan (1983) attempted to form a causal explanation of the correspondence relationship between worldview and subsisitence (hunting and distribution) in connection with the relationships between animals and humans. The stance of attempting to explain subsistence using worldview is shared by Stewart (1991) and Omura (2011, 2021). This worldview is a condition that unconsciously motivates people, whose primary role appears to be the function of explaining, ex post facto, success or failure in activities and convincing people (Irimoto 2007; Takekawa 2008).

16 Analyses of sharing must be conducted within the framework of a larger subsistence system as presented in a separate article (Kishigami 2008), while positioning cases in the context of globalization.

Chapter 4
Comparing Food Sharing in Three Hunting and Gathering Societies

4.1 Introduction

Hunter–gatherers have a strong tendency to share time, places, information, activities, and resources among themselves in a camp or settlement (Imamura 1993, 1996; Ingold 1988; Tanno 1991). They particularly give and receive food frequently. This transfer of food, designated as food sharing, is one characteristic of hunter–gatherer societies and/or small-scale societies (ex., Dowling 1968; Lee 1993; Terashima ed. 2021).

The forms of hunter–gatherers' food sharing can be categorized as giving, exchange, and redistribution. Until recently, anthropologists regarded all these forms of transfer of food into a single concept, i.e., food sharing (Kishigami 2003a, 2004a, 2007). In other words, cultural anthropologists treated sharing as a practice that took multiple forms. Since the end of the 1990s, however, many researchers have raised the opinion that hunter–gatherers' food sharing is not exchange. J. Woodburn (1998), for instance, argued that hunter–gatherers' food sharing is not exchange, but rather redistribution. Widlok (2017) regarded food sharing not as mutual exchange between a giver and receiver, but as a social practice by which the owner (giver) gives goods or services to the receiver based on the demands of the latter.

This chapter compares the forms, functions, and effects of food sharing in three hunter–gatherer groups including the Aka people in Africa, the Aché people in South America, and the Inuit people in North America to examine common aspects and differences in their food sharing among them. The reason for selecting these three groups for this study is that earlier studies about their food sharing were more detailed than those about other hunter–gatherer groups.[1]

N. Kishigami, *Food Sharing in Human Societies*, Trust 4,
https://doi.org/10.1007/978-981-16-7810-3_4

4.2 Framework for Food Sharing Comparison

Food sharing has been regarded as one characteristic of hunter–gatherer societies. Cultural anthropologists, however, have conducted research while leaving the meaning of "sharing" ambiguous rather than defining it clearly. It has been defined most recently by Widlok (2017),[2] who adopts the argument of Peterson (1993) that sharing is initiated by a demand of a receiver. Widlok defines sharing "as the social practice of enabling access to what is valued based on shared demands" (Widlok 2017: XVII). This definition emphasizes demands of the receiver rather than the desires of the giver. In other words, it is an act initiated by a person demanding goods such as food from another person who owns such goods, by which the receiver can obtain what is demanded from the giver, who gives up all or part of what is demanded. The perspective of this definition is original, but its effectiveness remains to be verified. The verification requires future testing in various cases worldwide.

This study provisionally uses a definition of food sharing in primatology, i.e., transfer of food from an individual to another not by force (Nishida and Hosaka 2001: 263–264), considering the current universality. Based on this definition, this study assesses the effectiveness of Widlok's definition of food sharing by comparing the three cases.

This chapter first explains several aspects of food sharing for comparison, i.e., the forms, functions, effects, and characteristics of food sharing. Subsequently, this chapter presents a description of food sharing practices found for the Aka people in Africa, the Aché in South America, and the Inuit in North America. It then compares food sharing in these three groups to elucidate similarities and differences of food sharing practices among them. It finally examines the definition of food sharing.

4.2.1 *Forms of Food Sharing*

Forms of food transfer, as noted earlier, include giving, exchange, and redistribution (Kishigami 2003b, 2004a, 2007: 30; Woodburn 1998). To explain these forms, we assume two individuals: A who has meat and B who has no meat.

Giving means that A one-sidedly gives all or part of his own meat to B. This case does not assume that B will return the meat or will give something in return to A in the future.

Exchange means that after A gives all or part of his meat to B, B returns some meat, other goods or services to A as a consideration for the meat or give something in return immediately or in the future. Therefore, food or/and services mutually move between A and B ($A \rightleftarrows B$) like an inter-flow: A—food → B, B—food or service → A.

Redistribution means that meat owned by A and other hunters is collected by another individual or organization acting as a center and all or part of such meat is then given to people, including A.

4.2.2 Functions and Effects of Food Sharing

Past studies of cultural and ecological anthropology have pointed out that food sharing has multiple functions and effects. This chapter presents discussion of five such hypotheses (Kishigami 2003a, 2008: 30–39). However, I do not directly address theories including kin selection based on evolutionary ecology, reciprocal altruism, tolerated theft, social signal, and cooperative labor because of difficulties in verification using ethnographic data.[3]

The first is the insurance function: diversification and reduction of risk. It is a view that food sharing is a means of resolving instability that prevails in acquiring food derived from hunting and gathering. This is known as a risk diversification function.

The second is a function of material and political leveling. It is an idea that hunter–gatherers can reduce disparities in the amounts of food owned by individuals and families through sharing hunted animals. At the same time, sharing has an effect of facilitating the maintenance of egalitarian relations in terms of political power. This is designated hereinafter as a leveling function.

The third is a function of reproducing and maintaining particular social relationships. Food is shared according to specific social relationships: with certain relatives, neighbors, and hunting participants. Food sharing practices visualize a particular social relationship between a giver and a receiver through frequent flows of food. By giving or receiving food, they affirm, confirm, and maintain a relationship between the giver and the receiver. It is also pointed out that social identity can be maintained and reproduced by having common awareness as relatives or associates among individuals who always or frequently share food. However, not sharing food according to existing social relationships signals a termination or relinquishment of such social relationships or a sense of belonging. Hereinafter, this is called a function of social relationship reproduction.

The fourth is a view that a skilled hunter who always shares a large amount of meat with others is able to improve his social reputation. It is noteworthy that this does not necessarily mean that such a hunter gains political power in his community. This is called the function of social reputation acquisition. This opinion closely resembles behavioral ecology's "show off" hypothesis.

The fifth is a function pointed out by Marcel Mauss[4] that food sharing creates, albeit temporarily, a relationship of superiority and inferiority between the giver and the receiver and the receiver's sense of indebtedness (Mauss 2014 (1924): 197, 295, 377). This is described hereinafter as a function of creating a superiority–inferiority relation and a sense of indebtedness.

4.2.3 Characteristics of Food Sharing

In addition to functions and effects, researchers in cultural and ecological anthropology have pointed out several characteristics of hunter–gatherers' food sharing. I will summarize several characteristics of food sharing that have been pointed out in the past (Kishigami 2003a; 2007: 30–39).

The first is that, whereas food is frequently shared in person in a small group, the sharing network stops functioning when the population becomes larger (Myers 1988; Kaplan and Hill 1985a, 1985b).

The second is that large amounts or pieces of meat tend to be shared widely. Small amounts or pieces of meat tend not to be shared (Graburn 1969: 66–67; Kaplan and Hill 1985a, 1985b).

The third is that hunter–gatherers' food sharing is connected to an environmental view of the relationship between people and the environment (Bird-David 1990) and a worldview of the relationship between people and animals (Bodenhorn 1990; Fienup-Riordan 1983; Nuttall 1991; Stairs and Wenzel 1992: 5). This view constitutes an ethical foundation for food sharing.

The fourth is that hunter–gatherers' food sharing is demand sharing initiated by a demand of the person wanting food (receiver) (Peterson 1993; Widlok 2017).

The degree to which the characteristics above can be generalized has yet to be verified. The cases of three groups in view of the forms, functions, effects, and characteristics of food sharing described above will be compared and examined later. Before that, food sharing as practiced by the Aka, Aché, and Inuit groups is described.

4.3 Food Sharing in the Aka Group

4.3.1 Subsistence Activity and Society of the Aka People

The Aka is a hunter–gatherer group living in the tropical rainforest of Central Africa. The people has been classified as a group once collectively called the Pygmy. Depending on regions and groups, Pygmy people vary in subsistence activity, residential group composition, and relations with neighboring farming societies and market economies to certain degrees.

I will specifically describe the Aka group living the area around the village of Linganga-Makaou, Dongou District, Likouala Region, and the Republic of the Congo, where Koichi Kitanishi conducted research in the 1990s. Approximately 350 Aka people reside near this village, constituting several residential groups comprising 3–20 related families (15–100 members) (Kitanishi 2004: 57). Families of a residential group unite and part from time to time. Therefore, their composition varies greatly through time.

This Aka group lives in forests for 4–8 months a year, moves to a nearby farming village, and stays there for the remainder of the year. When in a forest, they keep moving their camp every several days to several weeks while hunting and gathering food. When in a village, they assist in farmers' tasks, work odd jobs, exchange wild plants they collected with crops, and eat food such as cassava and other crops, meat obtained from hunting, and larvae. Although the Aka people of this region have contact with neighboring farmers, they are affected by outside farming communities and market economy less than other Pygmy groups in other regions.

Because the Aka people carry out hunting and gathering in the forest, their staple foods include wild yams, nuts, animal meat, honey, and crops (Kitanishi 2004: 60). The camp size varies from 3–6 families (15–20 members) to 13–17 families (approximately 60 members) (Kitanishi 2004: 68).

Women gather nuts and root vegetables such as wild yams. In view of caloric intake, gathered foods constitute a large part, including 17% from wild yams and 17% from nuts. Food procurement through gathering is stable. For greater numbers of family members, greater amounts of food can be gathered by each woman (Kitanishi 2004: 64). Men, by contrast, collect honey from the hives of honey bees and stingless bees. This comprises 15% of the total caloric intake (Kitanishi 2004: 60). The amount of honey collected varies considerably among individuals (Kitanishi 2004: 63).

Main hunting activities include trapping, spearing, and netting, which contribute approximately 26% to total caloric intake (Kitanishi 2004: 60). Trapping is individual hunting used to capture various animals: from small ones such as mongooses to large ones such as giant forest hogs weighing more than 100 kg. Many of the caught animals purportedly weigh more than 20 kg each (Kitanishi 1997: 9). Such trapping comprises more than 75% of the total amount of animals taken through hunting (Kitanishi 2004: 61). The animals hunted most frequently by spearing are bush pigs, each weighing 20–100 kg (Kitanishi 1997: 9). By netting, blue duikers weighing approximately five kilograms each are caught most frequently, but medium-sized duikers weighing approximately 15–25 kg are also captured (Kitanishi 1997: 9). The amounts of meat obtained from hunting per day vary considerably among individuals (Kitanishi 2004: 61).

Even in forest life, caloric intake from crops such as cassava obtained from farmers through exchange with wild plants and animals accounts for 22% of their total caloric intake.

A camp group comprises several families who share kinship relations. High social and political egalitarianism prevails within each group (Kitanishi 2004: 58–59).

4.3.2 *Food Sharing of the Aka*

According to Kitanishi (1997, 2002, 2004), the Aka people in the area of his research shared hunted animals at least three times. Reportedly, the occasions are divisible into, for instance, (1) sharing a hunted animal at the hunting site, (2) sharing it at a camp, and (3) sharing through distribution of cooked food at a camp. In sequence, the same animal meat is shared in multiple occasions, which can be designated respectively as the primary sharing, secondary sharing, and tertiary sharing.

4.3.2.1 Sharing of Meat

Trapping is hunting by individuals. No sharing occurs at a hunting site. The owner of a trap is the owner of the animal caught. The owner of the animal shares the meat voluntarily after returning to the camp.

In spearing, the primary sharing is conducted at the site of hunting or the camp. This sharing is conducted based on rules under which the second person who speared the animal receives the hips and the third person receives the head. If the person spears first and is not the owner of the spear, he is given the rump. Subsequently, the remainder of the animal is brought to the camp. The persons in possession of the animal meat voluntarily engage in secondary sharing to share the meat with other families.

In netting, primary sharing and secondary sharing are conducted at the hunting site. Tertiary sharing to distribute the meat to families whose members did not participate in the hunting is conducted at the camp. The owner of the net becomes the owner of the caught animal. In netting, the first person who captured an animal by the net is given the breast, belly, and intestines at the hunting site. The person who set the hunting net is given the rump (primary sharing). A woman who is a close relative of the net owner (normally his wife) butchers animals during net hunting and shares the meat with other women present at the site (secondary sharing). After returning to the camp, the meat owner and women who received meat voluntarily distribute their meat to women from other families in the camp (tertiary sharing) (Kitanishi 1997: 9).

Parts of the animal distributed in the primary sharing in hunting activities are specified by rules. The secondary sharing is conducted voluntarily by meat owners and people who obtained meat from the primary sharing. A piece of meat weighing approximately 10 kg or more is always shared with others. Approximately 70% of all meat is distributed (Kitanishi 2002: 74). They give the meat to more families at the camp if they have large amounts of meat to be shared. However, a small amount of meat is distributed to only a few families or is not distributed at all. In the secondary sharing of meat, the owner of meat obtained by hunting or primary sharing decides to whom and how much the owner will give the meat.

4.3.2.2 Sharing of Plant Foods

A woman, in some cases, shares plant foods with other women before cooking. Such cases include (1) when she returns home with a large amount of fruits, (2) when she has gathered a large amount of wild yams (more than five kilograms), and (3) when a large amount of crops such as cassava, its leaves, and oil palm fruits is brought from a village. Unlike meat, plant foods tend to be shared only with relatives (Kitanishi 1997: 12–13).

4.3.2.3 Sharing of Honey

An owner of honey is the person who has discovered a beehive. Honey, if it is collected, is consumed by people present at the site. Any remaining honey is brought to the camp and distributed through secondary sharing. Honey is given to all families of the entire camp, although close neighbors and relatives of the beehive owner in the camp tend to receive a larger amount than others.

4.3.2.4 Sharing of Cooked Food

Meat and edible plants that are obtained are generally transferred to a woman (wife or mother) who is the center of each family (household) in the camp. The woman makes a stew using the meat and plants of several types such as nuts. She then collects bowls from women from families to which she plans to distribute the stew, serves the food in the bowls, and provides them (Kitanishi 1997: 4, 2004: 66). Through this type of food sharing, the woman who has cooked the food decides at her discretion the families to whom the food will be provide and determines the amounts to be given, considering their kinship, spatial proximity, and other factors.

The greater the amount each woman cooks, the more families receive the food. The average is five bowls. One bowl is given to the husband and other men. Three are distributed to other families. The remaining one is taken by the woman and her children (Kitanishi 2004: 69). The men eat the bowl of stew at a gathering place called *mbanjo* at the center of the camp with other men who are present (Kitanishi 1997: 4, 2004: 66).

A small camp has 4–5 women (i.e., 4–5families) who are able to share food at all times (Kitanishi 2004: 69). The women at the camp make similar dishes in the late afternoon and share them with others on a daily basis (Kitanishi 2002: 89). She provides approximately 80% of the food she cooks to men including her husband and other families. She and her children consume approximately 20%. Therefore, a large part of the meals consumed by each family is cooked and provided by women to other families (Kitanishi 2002: 76). At a small camp, food is given to all families. Therefore, the dishes and the amounts consumed by the families are mostly equivalent (Kitanishi 2002: 89).

A large camp formed near a farming village has numerous families (population); one woman is therefore unable to distribute the food she cooks to all families there. Women therefore tend to share food with relatives living nearby (Kitanishi 2004: 74).

4.3.3 Characteristics of Food Sharing of the Aka

Several characteristics of food sharing of the Aka in their forest life can be explained as presented hereinafter. First, through several iterations of sharing of meat and dishes, all families and individuals living in the same camp obtain mostly the same types of food and dishes in nearly the same amounts. The sharing has the effect of leveling differences in the amounts of animals hunted among individuals (Kitanishi 1997: 12, 2004: 78).

Secondly, when a large amount of meat or plant food becomes available, food is provided to all camp members. When only a small amount is available, food might not be distributed (Kitanishi 1997: 11, 13). The amount of available food is apparently an important condition for sharing. Meat of bush pigs and blue duikers, which is valued higher than plant foods by the Aka, tends to be shared widely even when only a small amount is available (Kitanishi 1997: 22).

The third characteristic is that, whereas the primary sharing of hunted animals is conducted according to rules, secondary sharing and tertiary sharing are done at the discretion of the meat owners or women who cook the food (Kitanishi 1997, 2002: 70, 2004: 66).

The fourth characteristic is that food tends to be shared universally in small camps and selectively in large camps (Kitanishi 1997: 15). The condition for receiving food in a small camp is to be present at the time of sharing. The condition in a large camp is to have kinship or be spatially close to food owners (Kitanishi 1997: 19, 22).

The fifth trait is that food sharing creates and maintains a social relationship between the giver and the receiver. Many layers of close relationships are formed through many iterations of food sharing, creating a network of relationships. Food sharing becomes a means of maintaining such social relationships and of confirming them on a daily basis (Kitanishi 1997: 21, 2004: 85).

The sixth is that characteristics of the Aka's food sharing are based on rules and voluntary actions rather than on demand.

Finally, it must be emphasized that the characteristics described here are not always common among all Aka groups. Bahuchet (1990: 33), who researched the Aka group in Lobaye, pointed out that the secondary sharing of meat is based on kinship. Takeuchi (1995: 72), who surveyed the Aka people in areas near Moumpoutou village, stated that food was shared with people both who lived spatially close to them and who were closely related by blood. Such variation is regarded as a result of historical differences in subsistence activity, and political, economic, and social conditions among the groups.

4.4 Food Sharing in Aché Society

4.4.1 Subsistence Activity and Society of the Aché People

The Aché are hunter–gatherers living in lowland subtropical forests in the eastern region of Paraguay in South America. They were nomadic until about 1975. Now they have settled in a mission village established by a Christian organization. They make a living through economic activities that combine hunting, gathering, and slash-and-burn farming.

Research conducted in the 1980s suggests that, before settling down in a village, they formed band groups, each consisting of 15–28 members. They kept moving after staying only one to four days at each location. They individually hunted deer, armadillos, and collared peccaries and collectively hunted white-lipped peccaries, pacas belonging to the rodent family, and capuchin monkeys. They also collected honey for food. Such animals and honey were procured mostly by men, who were engaged in hunting for six hours a day, on average. Each hunter reportedly acquired approximately four kilograms of meat during a day's hunting. The probability of not being able to hunt any animal was approximately 40%. Approximately 60% of captured animals were caught during group hunting. Hunting skills clearly varied among the male hunters. Women gathered fruits, insect larvae, and palms containing starch (starchy palm fiber) (Gurven et al. 2002: 96). In terms of calorie intake, 60–80% was derived from the meat of hunted animals, followed by honey, fruits, plant foods, and insect larvae (Kaplan et al.1984: 113).

In 1978, the Aché in the northern region settled in a Catholic mission village called Chupa Pou. One group that settled there then left the village. In 1980, they established a mission village called Arroyo Bandera on the advice of a Protestant missionary group (Gurven et al. 2002: 96). In 1998, 117 people from 23 nuclear families resided in Arroyo Bandera. Each household (nuclear family) had built a wooden house with an area of approximately 16 square meters, located approximately 100 m distant from one another. The Aché people began to grow cassava, pulses, peanuts, corn, sweet potato, and sugarcane in small fields and raise fowl, hogs, cattle, horses, pet monkeys, coatis, and peccaries. Additionally, they worked at farms and ranches operated by Paraguayans and used the wages to purchase clothing, tea, salt, sugar, bread, and pasta. They occasionally made palm leaf fans, mats, bows and arrows, and other crafts to sell to tourists and missionaries. Meals at the mission village comprised approximately 80% of crops, approximately 9% of foodstuffs purchased at stores, and approximately 11% of meat of wild animals and livestock. Aché people in the village often engaged in collaborative work such as farm work and livestock farming, using mutual assistance and borrowing and lending tools among themselves (Gurven et al. 2002: 97).

As of 1998, the Aché based in the Arroyo Bandera village spent approximately 20% of their time in the forest, hunting and gathering. In a forest camp, they made 3–6 open-air fires in a space with a radius of 3–5 m. There, 3–10 nuclear families erected temporary huts around the fires.

Kaplan and others accompanied hunting and gathering forays into forests nine times during October 1981 through May 1982 and conducted research on hunting and gathering activities and food sharing for 81 days (Kaplan and Hill 1985; Kaplan et al. 1984). M. Gurven and others studied food sharing in the Arroyo Bandera village during February–May 1998 (Gurven et al. 2002). The following describes their research results. Their investigations were undertaken to verify the importance of food sharing using data quantified based on a hypothesis derived from evolutionary ecology. It is noteworthy that they lacked qualitative descriptions of ways, forms, and other factors of food sharing among the groups involved.

4.4.2 Food Sharing of the Aché

4.4.2.1 Food Sharing in a Forest Camp

The study of Kaplan and others did not distinguish information related to sharing at the time of hunting and gathering, sharing after returning to a camp, and sharing of cooked food, but quantified all information in aggregate (Kaplan and Hill 1985a, 1985b; Kaplan et al. 1984). The information was organized to the greatest degree possible to facilitate comparison with results obtained from other studies.

Whereas 3–10 nuclear families often stayed together in a camp in the forest, the unit of food consumption was the nuclear family. Food obtained from hunting and gathering was shared frequently in the camp. The Aché people would consume food within a day of procuring it and would rarely store it. Kaplan and others summarized this characteristic as described below (Kaplan et al. 1984: 114).

First, an Aché person received a large part of their food, in terms of caloric intake, from people outside their nuclear family. Because the skills of hunters varied in hunting, there were always people who caught numerous animals and those who did not. When successfully acquiring an animal, hunters shared much of its meat with others. Whereas hunters shared the animals they captured with people outside their own nuclear family, they and members of their nuclear families received animals from hunters of other nuclear families (Kaplan and Hill 1985a: 229).

Secondly, food shared in a camp in forests varied depending on the type. Food that was distributed most frequently was meat, followed by gathered foods such as honey and fruits, which was followed by food brought from mission villages (Kaplan and Hill 1985a: 229). Differences in distribution patterns were observed among these types of food. Whereas meat and honey were shared among hunters and nuclear families in the camp, food brought from mission villages tended to be shared and consumed among members of the nuclear family. In other words, meat and honey were often shared with other nuclear families in the camp and were not necessarily provided more to specific close relatives (Kaplan and Hill 1985a: 235). Food of other types such as fruits and root crops, however, was gathered by many people simultaneously and was provided more to parents, children, siblings, spouses, and other close relatives. Therefore, food of these types would not be shared with people outside nuclear families unless very large amounts were to be gathered.

Thirdly, the amounts and frequencies of sharing meat among relatives in a camp were mostly the same as sharing with people not related by blood (Kaplan and Hill 1985a: 229, 235). Sharing of meat in a hunting area or a camp in forests was affected by factors other than kinship, such as engaging in the same hunting and being at the same place.

Finally, in a forest setting, people in a camp were able to procure food constantly through food sharing and were able to maintain a better nutrition condition (Kaplan and Hill 1985a: 236).

4.4.2.2 Food Sharing in a Mission Village and Forest Camp

This section presents a description of food sharing of the Aché in the Arroyo Bandera village and neighboring forests at the end of the twentieth century. At the time, the Aché people would go hunting and gathering from their village and return to the village thereafter. They spent approximately 80% of their time in the village and 20% of their time in the forests during the year.

Life based in a village brought changes to hunting and gathering activities of the Aché (Gurven et al. 2002: 96). First, a hunter eating the meat of an animal that he had caught and killed himself was considered taboo in the old days. This had since ceased to be taboo. Secondly, they changed their hunted animals. Although the numbers of hunted white-lipped peccaries and coatis decreased, other animals such as armadillos, pacas, and capuchin monkeys began to be caught in greater numbers. Finally, the sizes of camps (housing) in forests and the average periods of hunting and gathering trips decreased.

Food Sharing in a Forest Camp

This section presents a description of food sharing in a forest. When hunting and gathering in a forest, hunters caught animals (5.8 kg on average) and collected honey (6.5 kg on average); then they returned to their camp and widely shared the food with other people (Gurven et al. 2002: 99). Approximately 80% of the food acquired through hunting and gathering was shared with other nuclear families in the same camp. More specifically, hunters and gatherers provided 89% of meat, 87% of honey, 73% of palm sprouts, 70% of palm starch, 70% of fruits, and 59% of insect larvae that they obtained to other nuclear families.

Meat and honey, which were widely available in the forest, tended to be provided frequently to many families in a camp (Gurven et al. 2002: 104). The amount of food acquired by each nuclear family was given to an average of 2.6 other nuclear families or approximately 41% of the residents of the entire camp (Gurven et al. 2002: 104). Each nuclear family reportedly received approximately 56–84% of the food which had been hunted and gathered by other nuclear families (Gurven et al. 2002: 105). As suggested in this case, each nuclear family gave more than a half of the food it had obtained to other nuclear families and received more than a half of its own food from other nuclear families.

As described earlier, hunting skills varied among hunters. Therefore, in some cases, only some hunters caught animals. Other hunters might be unable to catch any animal. As a result, there were some nuclear families which gave more food to others and other nuclear families which received more food from others in a camp. Food sharing presumably leveled the food intake of nuclear families in the camp.

Food Sharing in a Mission Village

Food sharing was practiced in a mission village of 117 residents (23 nuclear families). Each family gave 78% of crops such as cassava, pulses, corn, and potatoes, 91% of livestock meat such as chicken, beef, and donkeys, 90% of meat of animals brought from the forest to the village such as armadillos, pacas, monkeys, tapirs, and peccaries, 91% of forest products such as honey, insect larvae, and fruits, and 75% of food purchased at stores such as bread, salt, and sugar to other nuclear families (Gurven et al. 2002: 99).

In the mission village, food obtained by a nuclear family was distributed to an average of 2.1 other nuclear families (approximately 9% of the total number of nuclear families in the entire village). Livestock meat such as donkey meat (65 kg) and beef (110 kg) was given to an average of 4.4 other nuclear families. Similarly, animal meat brought from the forest to the village was distributed to 2.3 nuclear families, food brought from other forests was distributed to 2.7 nuclear families, crops were provided to 1.6 nuclear families, and food purchased at stores was given to 3 nuclear families (Gurven et al. 2002: 105).

In the mission village, foods of almost all types were given to close relatives in larger amounts and more frequently than others. The residential locations of nuclear families practicing food sharing, i.e., distances among houses and whether the houses of other families were visible, also affected the sharing practices. Then they tended to share food with nuclear families living physically close to them (Gurven et al. 2002: 111). In other words, kinship and physical distances were important conditions affecting the amounts and frequency of food sharing in the mission village (Gurven et al. 2002: 112).

4.4.2.3 Continuity and Differences in Food Sharing in the Forest Camp and Mission Village

Both differences and common characteristics were observed in food sharing between the forest camp and the mission village (Gurven et al. 2002: 113).

Food sharing was practiced frequently both in the camp and in the mission village. In both cases, approximately 80% of the total weight of food procured by a nuclear family was shared with two or three other nuclear families.

Physical distances among nuclear families practicing food sharing were important both in the camp and mission village. In both cases, food sharing was practiced more frequently among nuclear families living nearby than among others.

Differences in food sharing between the forest camp and mission village are the following. In the forest camp, nearly all animals hunted and food gathered were shared among nuclear families, but 67% of the nuclear families in the mission village shared food with other nuclear families. In both communities, physical distances among nuclear families were important factors to be considered when deciding how much food would be given to whom. In the mission village, kinship became an important factor in deciding with whom food should be shared. A small amount of food could be distributed to all nuclear families when the number of nuclear families living in the same area was small. It became difficult, however, to provide food to all nuclear families when the number of nuclear families was large. Therefore, the giver had to select receivers. In other words, being at the same place (physical proximity) was an important factor in food sharing in a small camp, whereas both physical proximity among houses and kinship were important in food sharing in a larger village or camp.

4.4.3 Characteristics of Food Sharing of the Aché

This section presents a summary of the characteristics of food sharing of the Aché based on the studies described above, which were conducted at the beginning of the 1980s and at the end of the 1990s.

First, food sharing is practiced frequently both in the camp and mission village. Types of food shared and recipients of the food, however, differ between them. Although physical proximity, i.e., being at the same place, is important in the camp, physical proximity, i.e., distances among houses, is also important in the village. Additionally, kinship is important in the mission village.

Secondly, more than half of the food acquired by each nuclear family is shared with members of other nuclear families.

Thirdly, foods such as animal meat and honey that are amply available tend to be distributed to more families or people, whereas small amounts of food are shared only with specific recipients or are not shared at all.

Finally, food sharing presumably levels the food intake of nuclear families in the camp in the forest.

Because Kaplan and others provide very little information of social relationships related to food sharing, one can only make assumptions, but food sharing in the mission village having a large population must be selective in its choice of recipients when only a limited amount of food is available. Apparently, this is the reason for the strong tendency of sharing food among relatives and the clear presence of the function of confirming and maintaining existing kin relationships. It is noteworthy also that an absence of information about whether a receiver's demand is a characteristic of food sharing prevents any conclusion in this regard.

4.5 Food Sharing in the Inuit Society

4.5.1 Subsistence Activity and Society of the Inuit People

Inuit are hunting and fishing people inhabiting the Arctic region of Canada. Their direct ancestors were whaling people who migrated from Alaska to the Canadian Arctic about 1000 years ago. Because of global cooling, they shifted their central subsistence activity from bowhead whale hunting to procurement of seals, caribou (wild reindeer), and arctic char. Moreover, they started a lifestyle of seasonal migration. Subsequently, they began sporadically contacting non-indigenous whalers, cod fishermen, explorers, and subsequently, missionaries who began arriving from Europe to the eastern Arctic region of Canada from approximately the 1500s.

An important change was the Inuit's eager participation in the trading of arctic fox furs in Arctic Canada in the 1920s. Around the 1960s, they settled in villages established by the federal government of Canada, from which they went hunting and fishing and to which they would return thereafter. They make a living in the mixed economy of subsistence hunting–fishing and wage labor. At present, many Inuit people have jobs in villages and engage in wage labor. They hunt and fish in their spare time or on holidays. Whereas they go camping for more than two weeks during the summer, the periods of summer camps have decreased as time goes by after beginning of their life in the villages. People who do not go camping have been increasing. Hunting and fishing began to be conducted by individuals and small groups of people using snowmobiles, canoes with outboard motors, four-wheeled vehicles, high-powered rifles, and fishing nets. In fact, group hunting has become very rare except for beluga whale and walrus hunting. The people, however, frequently share food from hunting and fishing at summer camps and in villages.

The village of Akulivik, where I conducted fieldwork during 1984–2016, is located on the northeastern coast of Hudson Bay (latitude 60° 48′ N, longitude 78° 12′ W). In the current village, prefabricated houses were built in an area of 2 km east and west and 500 m north and south inland from Hudson Bay. In 2016, the total population of the village was 632; the total number of households was 155. The average number of residents per household was therefore 4.1. The Inuit people of the village obtained food by purchasing products from the southern region of Canada at stores in the village and by hunting and fishing. Approximately 60% of their protein intake was from local meat and fish. More than 60% of their total caloric intake was from food and foodstuffs such as bread, pasta, potato, pizza, sausage, ham, eggs, beef, pork, chicken, flour, lard, butter, milk, and sugar purchased at stores. In other words, the Inuit got approximately 60% of protein from local meat and fish and more than 60% of total calories from food products from outside their communities. Whereas people in elderly households tended to prefer locally produced food, people in young households tended to prefer Western food (Kishigami 2005b).

4.5.2 Food Sharing of the Inuit

The food sharing of the Inuit can be classified broadly into sharing at the time of hunting and sharing in a village. The forms of food sharing of the Inuit show regional differences and historical changes (Damas 1972a; Fienup-Riordan 1983; Kishigami 1998, 2000, 2004a, 2004c, 2007; Nuttall 1991; Ready and Power 2018; Wenzel 2000). In the region of Nunavik, village-wide country food distribution via the hunter support program began in the 1980s. This chapter uses data from research I conducted in the village of Akulivik in the Arctic region of Quebec Province during 1984–2004 (Kishigami 1998, 2007, 2010).

4.5.2.1 Primary Sharing of Hunted Animals at Hunting Sites

When multiple hunters caught an animal such as a seal, caribou, or walrus, each animal became the possession of the hunter who caught it. A part of it was shared with people who participated in its hunting or who were present at the butchering site. Generally, the hunter who caught the animal took a slightly larger portion. Then people present at the hunting or butchering site took a small piece each from the remaining meat. Participation in the hunt or being at the butchering site was an important condition for this distribution, irrespective of special relationships such as kinship. If individuals other than the hunter considered that the amount of meat was insufficient for distribution for them, then they might relinquish the sharing voluntarily.

When a caribou or a large amount of arctic char was caught by several hunters working cooperatively, the catch might be divided among the participating hunters. In this case, too, participation in the hunting or being at the butchering or sharing site was an important condition, irrespective of the kin relationships.

In the case of Inuit, the part of the hunted animal taken by the hunter who caught it and other parts given to others were not specified by rules in the primary sharing.[5] In addition, recent hunting partners were often members of an extended family such as a father and his son, siblings, cousins, and a grandfather and his grandchild. Naturally, the catch was often shared among hunters of the same extended family living in different households.

4.5.2.2 Food Sharing at Hunting and Fishing Camps

When captured animals were brought to a hunting or fishing camp, the meat or fish might be shared with people who did not participate in the hunting or fishing, and family members. At a summer camp comprising several families, the caught animals were shared by eating the meat together (participating in meals). Although the entire camp often consisted of an extended family, people not related by blood or marriage who were temporarily staying in or visiting the same camp also shared meals.

4.5.2.3 Food Sharing in a Village

After bringing a hunted animal to the hunter's home village, the hunter often kept the meat at home or took it to the home of his parents or grandparents. Children and grandchildren visited their parents and grandparents almost every day to have lunch or dinner and ate the meat together, except at breakfast. Meat of wild animals and fish was shared on a daily basis among extended family members, who often ate them together. People without food to eat might be able to obtain food by joining lunch or supper at the households of their friends or neighbors.

When a village population exceeded 50 or 60 (the population of Akulivik village was over 500 in the 1990s), it became physically impossible for a hunter to share his game meat or fish with all people in the village. Therefore, the daily sharing of meat and fish tended clearly to be conducted within each extended family or with close neighbors.

Hunters sometimes delivered meat and fish to widows, elderly people, and families without hunters. Villagers who ran out of food at home sometimes visited hunters asking for meat or fish. This was a kind of demand sharing in a broad sense. In this case, food was shared at the discretion of the hunter or his wife. Such sharing was also practiced for people who are not related genealogically.

At special occasions such as Christmas and Easter, community feasts for the entire village were organized and held by the village council or church committee at a gym in the village. These feasts were joined by most of the villagers.

4.5.2.4 Food Sharing Through the Hunter Support Program

As people have increasingly spent more time in a village since the 1960s, the frequency of their hunting and fishing has been decreasing annually. The summer camping period has also been reduced. Therefore, the Kativik regional government of the Nunavik region in Quebec province introduced a hunter support program in the mid-1980s after negotiating with the provincial government of Quebec in an effort to maintain and promote hunting and fishing. Using this program, Akulivik village began to encourage its hunters to go to remote regions using large community boats to hunt beluga whales and walruses, and provide the animals to the entire village. Through the new sharing program, they began to be able to provide meat of animals that each Inuit individual could not have caught alone (Kishigami 1998, 2000, 2007, 2010).

The village council purchased animals such as caribou and arctic char from hunters in the village using this program, stored the meat and fish in a freezer owned communally by the village, and provided the meat free of charge as necessary to Inuit people in need of food (Celsa 2002; Gombay 2005a; Kishigami 1998, 2007, 2000, 2010).

4.5.3 Characteristics of Food Sharing of the Inuit

The Inuit today subsist based on an economic system that combines their traditional economy and a monetary economy. Their frequency of hunting and fishing has been decreasing gradually. Food sharing of animal meat and fish provided by hunting and fishing continues to be practiced, albeit less frequently. The subsequent section presents a summary of the characteristics of food sharing of the Inuit.

First, no rule exists for sharing an animal captured in the first sharing at a hunting site.[6] The hunter who caught an animal or fish takes a slightly larger portion of the meat or fish; the remainder is allocated to participants in the hunting or fishing or people present at the butchering site. Meat or fish is often shared equally when a large amount is available.

Secondly, food is shared primarily by eating together in a camp or village. In a small camp consisting of several families, all people staying in the same camp tend to participate in meals, irrespective of their blood relationships. In a village with several hundred residents, people often join lunch or dinner at the homes of their parents, grandparents, uncles, or aunts and eat together mostly within each extended family.

Thirdly, several food sharing practices in Akulivk are characterized by a form of redistribution. For example, sharing through the hunter support program administered by village council occasionally provides beluga *maktak*, walrus meat, caribou meat, and fish to all Inuit households in the village. In addition, the church committee organizes community feasts at Christmas and Easter. In those cases, food is distributed to villagers by village council or church committee after obtaining food from hunters and shops.

The fourth characteristic is that food sharing has the function of leveling food supplies at the individual or family level. Hunting and fishing skills vary among individual hunters. Whereas some successful hunters or families often give food one-sidedly, other hunters or families lacking food often receive it one-sidedly in the village. Nearly all Inuit people, however, are able to obtain food through food sharing, which also has a leveling function that reduces differences in access to food. Food sharing of the Inuit, however, is most often practiced within each extended family. Disparities are reduced among individuals or households within each extended family. Whereas food sharing beyond the boundaries of extended families is practiced in the village, the frequency of that activity is lower than food sharing within an extended family. Food sharing in an Inuit village is therefore regarded as having the function of securing and leveling food supplies within each extended family.

Fifth, presence at the same place and kinship are important factors in the food sharing of the Inuit.

Sixth, food sharing practiced among extended family members and Akulivik residents contributes to confirmation and maintenance of specific social relationships including those based on kinship, friendship, being co-villagers, etc. in the village.

Additionally, communal meals at the village level have the function of confirming and maintaining the villagers' collective identity or we-feelings.

Seventh, the current practice of country food sharing contributes to the procurement of culturally valued food. The food consumption provides the Inuit with cultural satisfaction.

Eighth, food sharing of the Inuit is generally based on the voluntary act or rules of the giver. Inuit people in need of food sometimes visit others asking for food; alternatively, they join a lunch or supper. Although this action is demand-sharing in a broad sense, voluntary sharing and rule-based sharing are practiced more frequently among the Inuit.

4.6 Comparative Analysis

This chapter has described food sharing practices in three hunter–gatherer groups living in different regions and having different political and economic conditions approximately in the 1990s to early 2000s. I will compare these cases in view of food sharing forms, functions and effects, and characteristics.

4.6.1 Forms of Food Sharing

Earlier reports have described studies of the forms of food sharing studied by cultural anthropologists in the past included giving, exchange, and redistribution (Kishigami 2003b, 2004a, 2004c, 2007).

In the Aka group, much of the primary sharing, secondary sharing, and tertiary sharing of animals and food acquired is practiced based on rules or voluntary transfer of the food owner. Once a woman who plays the central role in each nuclear family obtains or receives meat and edible plants, she cooks and provides dishes to other nuclear families in a camp. The sharing of cooked food in the camp can be regarded as a form of redistribution.

In the study of the Aché group, Kaplan, Gurven, and other researchers specifically examined food sharing on the assumption of research that it is reciprocal and that it is regarded as an exchange. Although they statistically understood the flow of food in view of weight and calorie intake, they did not describe the forms in detail. The forms of food sharing in the Aché community are therefore unknown.

In the Inuit community, the basic form of sharing animals captured is voluntary giving. They share food by joining lunch and supper at the homes of their parents and grandparents. The sharing of meat and fish through the hunter support program takes the form of redistribution because collected such food in a central office of the village is distributed to villagers by the village officials. This is regarded as a form of redistribution.

The forms of food sharing in the three communities suggest that the basic form is likely to be giving and redistribution.

4.6.2 *Functions and Effects of Food Sharing*

4.6.2.1 Function of Risk Diversification

Hunting activities in the Aka, Aché, and Inuit groups are often affected by individual skills and luck. In all three communities, skilled hunters are reportedly the principal providers of animal meat and other food. Unskilled hunters become the receivers in many occasions. In the three groups, there are individuals who have sufficient food to provide for other camp members and relatives who do not, through food sharing which all families gain access to food. Food sharing can therefore be viewed as serving the function of diversifying risks in the people's food procurement. Food sharing, thereby, has a function of ensuring food security to certain degree.

4.6.2.2 Leveling Function

Because food sharing in the three groups is a means for successful hunters and their families to provide meat and other foods to other individuals and families, leveling of access to food and consumption among individuals and families has been observed. However, it should be emphasized that not every member in a camp or village receives and eats equal amounts of food.

The daily sharing of food and daily lunch and supper in the Inuit village have a function of leveling food consumption within each extended family. Sharing of meat and cooked food in the Aka and Aché villages is also regarded as having the function of leveling food consumption among individuals in each extended family and among family households in a camp or village.

Whereas leveling of the amounts of food possessed and consumed by food sharing is observed in these three groups, clear differences exist between camps with small populations and villages with large populations. In other words, in the former, sharing among all families (households) is possible, allowing equalization. In the latter, distributing a limited amount of food to all households is difficult, resulting in distribution within specific extended families (relatives) or groups. Consequently, leveling of food possession and consumption within each extended family occurs, but not across all families.

4.6.2.3 Function of Reproduction of Social Relationships

In the Aka, Aché, and Inuit groups, captured animals and food are shared among individuals and family groups comprising a camp. The social relationships of camp members are affirmed and maintained through visible food transfers between or among particular persons by sharing.

In food sharing in a village with a larger population, providing food to numerous families and people is difficult because of the limited amounts of food. Therefore, people give food primarily to close neighbors, immediate family members, or extended family members. Food sharing within a village is therefore thought to have a function of confirming and rebuilding social relationships based on neighboring and kin relationships.

4.6.2.4 Function of Acquiring Social Reputation

The function of food sharing to acquire social reputation was not verified in the Aché community because of data inadequacy. Whereas the *tuma*, a master elephant hunter of the Aka, receives social reputation, hunters' social recognition through food sharing could not be verified among the Aka. In the Inuit group, people who provide food to other people and families presumably gain good reputation. Such a reputation, however, is not necessarily connected to political power in the village, which is seemingly a characteristic of hunter–gatherers.

Verification of the relationship between food sharing and reputation will be left for future studies. I consider one reason for the continuous giving of food by Inuit hunters is the psychological satisfaction gained from a good reputation awarded by other community members (villagers) (Kishigami 2007: 147, 257). Hunters in the Arctic region also presumably think that it is important for them to contribute to the well-being of the entire community and find joy in practicing food sharing in itself (Bodenhorn 2000; Fienup-Riordan 1983; Kishigami 2007, 2012, 2014; Wenzel 1991).

4.6.2.5 Function of Creating a Superiority–Inferiority Relation and Sense of Indebtedness

Mauss (1924 (2014): 197, 295, 377) pointed out that one-sided gifting of goods and food would create a relationship of superiority and inferiority between the giver and the receiver and the sense of indebtedness in the receiver. Skilled hunters in the three groups one-sidedly provide less skilled hunters and their families with more food. No evidence exists for any of the three, however, that less skilled hunters and people who are always given food are placed in an inferior position and carry the sense of indebtedness. As they are taught from infancy, they come to think that it is natural for people who have food and captured animals to share them with those who do not. Consequently, they think it is culturally natural to seek food from those who have it when they do not. This prevailing set of practices and thoughts suggests, in other words, that food sharing practices in hunter–gatherer communities do not necessarily create superiority or inferiority or a sense of indebtedness.[7]

The reason, hypothetically speaking, is that elders such as parents, grandparents, and uncles and aunts continue to teach their youngsters from their infancy that food sharing is part of a social norm and a ordinary behavior. Normally, when questioning people who practice food sharing why they give their animals they hunted or food to others, most of them do not even understand the meaning of the question and are not able to express the reasons in words. The reason is that, in the process of growing to adulthood, food sharing is taught indirectly as an adequate practice that is not to be questioned and which is internalized unconsciously (Hewlett 2020: 112, 129–130; Kishigami 2016: 10; Sonoda 2019: 249). In this condition, food sharing rarely creates superiority and inferiority between the giver and the receiver or a sense of indebtedness in the receiver.

One surprising experience from my research among the Inuit in the mid-1980s was that receivers would not express appreciation such as "thank you" when provided with meat by other hunters. The attitude and mood suggest that it was natural for the receiver to be given the food.[8] However, persons who always asked for food from others without giving at all were spoken ill of or had a bad reputation. From my experience during fieldwork, food sharing was seemingly an ordinary activity at the time and did not appear to create a difference in social status between the giver and the receiver. For clarification, it is noteworthy that in the current social context, food sharing among non-relatives might create a superiority–inferiority relation between food givers and receivers and the food receivers might express appreciation to the givers because of a decrease in the amount of meat to be shared and frequency of food sharing caused by rapid social and economic changes in Inuit society.

4.6.3 Characteristics of Food Sharing in Hunter–Gatherer Groups

This section presents a discussion of food sharing in hunter–gatherer groups in relation to four characteristics, including population size, amount of food, environmental views and worldviews, and demand.

4.6.3.1 Population and Food Sharing

Food sharing is known to involve all people in a camp group of hunter–gatherers comprising up to about 30 members from several families (Kelley 1995: 211). Winterhalder (1986a, 1986b) pointed out that the effect of food sharing would be maximized in a small group with seven to eight hunters. Smith (1988) reported that when a group grew larger and communication was less personal, people would start storing food and food sharing would no longer function effectively. Continuous practices of food sharing require an ability of mutual monitoring, which, however, becomes more difficult as the group grows larger (Smith 1988).

The studies revealed differences existing in details of food sharing between camps with a small population and villages with a large population in all Aka, Aché, and Inuit societies. In the former, food sharing was conducted frequently. It involved all residents of mostly the same camp. In the latter, at least the receivers of food were selected. Sharing was more limited. Particularly, factors common in all three groups included that, in the former, sharing the place, i.e., being in the same camp, was important irrespective of kinship and that, in the village, food was given in larger amounts and with higher frequency to close relatives of successful hunters, which indicates that, in hunter–gatherer groups, sharing the place by a residential group with a small population and kinship for a group with a large population are important conditions for food sharing practices.

4.6.3.2 Food Amount and Food Sharing

Kitanishi (1997: 11, 13), who conducted research among the Aka people, remarked that they tended to share animal meat in large amounts with more people (or families) in a camp and plants and other food gathered in small amounts with fewer people (or families). A similar remark was made based on research conducted to investigate the Aché people (Gurven et al. 2002: 104). In the Inuit society, too, animals such as beluga whales and walruses far exceeding 500 kg each or large catches such as arctic char captured in a large number tend to be shared widely beyond extended families in both summer camps and a village. Berries gathered in the summer are rarely shared but are often consumed by the individuals who gathered them, and within their families.

These findings suggest that in hunter–gatherer societies, animal meat in larger amounts tends to be shared more widely. Food in a smaller amount or a small animal hunted, however, tends not to be shared.

4.6.3.3 Environmental Views and Worldviews and Food Sharing

Researchers have expressed that the ideas of environmental views and animal views held by hunter–gatherers are related to food sharing (Bird-David 1990; Bodenhorn 2000; Brewster 2004; Fienup-Riordan 1983, 1999; Kishigami 2012, 2014; Nuttall 1991; Omura 2021; Stairs and Wenzel 1992; Stewart 1991). Bird-David, who has studied the Nayaka, hunter–gatherers in India, for instance, stated that the Nayaka people considered the forest environment to be like a parent providing them with food. Bird-David (1990) argued that this environmental view was the ethical basis by which the Nayaka share their food with others. The Iñupiat in Alaska believe that because bowhead whales willingly surrender their life to hunters and get hunted, hunters should not keep the meat and other edible parts to themselves, but share them with other people (Bodenhorn 2000; Kishigami 2012, 2014).

Although Kitanishi, introduced in this chapter, did not report the environmental views of the Aka people, Turnbull, who investigated the Mbuti people, wrote that the Aka people viewed forests as their father and mother, providing the people with food and other benefits. Hewlett (2020: 79–80, 142–143) argued that the Aka's environmental views and their relation with the environment were profoundly related to their values, which had become the basis of food sharing. The relationship of the Aché people's food sharing and their environmental views has yet to be reported. For the Inuit, the relationship between their views of animals and sharing of caribou meat and beluga *maktak* has been reported (Stewart 1991; Kishigami 2000). Indigenous people in the sub-arctic region of North America also show such a relationship (Nadasdy 2012).

Although this relation between food sharing and views on the environment and animals might apply to all three groups, clear evidence for the Aché people does not exist, which requires further investigation.

4.6.3.4 Food Sharing Based on Demand

Are the basic characteristics of food sharing of hunter–gatherers regarded as based on demand ("demand sharing" hereinafter)? This question is related to the validity of the definition of sharing newly proposed by Widlok (2017).

The concept of demand sharing specifically emphasizes "demand" that appears among Australian Aboriginals as the trigger for the initiation of food sharing. Peterson (1993) conceives that this form of sharing applies to nearly all hunter–gatherer societies. Two factors are noteworthy in this case. One is that demands of a person without food for food from a person with food occur in nearly all hunter–gatherer societies, which is, in this sense, universal. The other is that the forms and intensity of expressing the "demand" vary, including pleading, strongly demanding in words, and implicitly requesting by appearing at a place and time of eating (Maruyama 2016). In a broad sense, demand sharing is regarded as practiced by nearly all hunter–gatherer groups. Whether demand sharing is the mainstream form of food sharing among hunter–gatherers, however, is another issue.

In food sharing in the Aka group, the sharing of cooked food is voluntary. The primary sharing of a hunted animal is specified by rules, suggesting that demand sharing would not necessarily be the mainstream even if it were practiced. No conclusion can be drawn for the Aché group because of scarcity of information. In the Inuit group, voluntary and rule-based sharing is plausibly the mainstream whereas giving in a form such as demand sharing is certainly practiced. Based on the above, I argue that demand sharing is not necessarily a basic form of sharing in all hunter–gatherer groups, even if it is practiced widely.

Widlok initially thought "sharing is allowing others to take what is valued" (Widlok 2013). Then this idea was developed to imply that sharing is a social practice that would allow access to valuable goods and services based on demand sharing (Widlok 2017: XVII). Based on cases presented in this chapter, it is difficult to conclude that "demand" is a core and fundamentally important aspect of food sharing. A limit might exist to the validity of the definition of "sharing" based on demand sharing.

Based on the view[9] presented by Belk (2010: 717), who compares and analyzes the concepts of sharing and gifting, I argue that sharing be regarded as a social practice that transfers resources such as food from an individual to another not by force for people to access useful resources or transfer useful resources to others, which I believe is more useful as a concept for description and analysis.

4.7 Conclusion

In hunter–gatherer societies, hunters often give their meat to others or share food with others. Such acts have been designated collectively as "food sharing," of which the forms and characteristics, however, vary among groups, regions, and time periods.

This chapter has presented a comparison of the forms, functions and effects, and characteristics of food sharing in the Aka, Aché, and Inuit groups in an attempt to identify common aspects among them. Following are the results.

The basic forms of food sharing of hunter–gatherers are giving and redistribution. This view does not mean that the form of exchange does not exist in hunter–gatherer groups. Rather, it implies that food sharing is rarely practiced as the form of exchange in the groups.

The functions and effects of food sharing in hunter–gatherer groups are listed below. First, food sharing has a function of diversifying risks in the people's food procurement. Secondly, food sharing has the effect of leveling the amounts of food in temporary possession and amounts of consumption among individuals and families. Thirdly, social ties among specific individuals and families within a community are confirmed and maintained by practicing food sharing. The fourth effect is the likelihood of gaining social reputation by individuals who often share food with others. In the Aka and Aché groups, however, clear evidence of this was unobtainable from the existing research results. Finally, unlike the remarks of Mauss in the gift theory, food sharing in hunter–gatherer societies does not necessarily create a superior–inferior relationship and a sense of indebtedness. The explanation presented above suggests that the practice of food sharing in hunter–gatherer groups has multiple functions and effects (Kishigami 2003a, 2003b).

The following are results of comparing the characteristics of food in hunter–gatherer groups. First, characteristics of food sharing vary depending on the size of the group among which food sharing is conducted. In hunter–gatherer groups, when the population of a camp group is small, "sharing a place," including proximity and co-presence, is an important condition or factor in practicing food sharing. When the population is large, kin relationships such those of parent and child and siblings become important conditions or factors. Secondly, hunted or gathered food of a large quantity is distributed more widely among hunter–gatherer groups. Thirdly, a profound relationship apparently exists between food sharing and the people's views of the environment and animals. Generalization, however, is limited because of the absence of clear evidence of such a relationship in the Aché society. The fourth characteristic is that demand of those in need of food for food from those in possession of food is observed in nearly all hunter–gatherer societies. This demand sharing, however, is not necessarily a key characteristic of food sharing. This fourth conclusion demonstrates the limited validity of Widlok's definition of "sharing" based on a concept of demand sharing. I propose the definition that sharing is a social practice that moves resources such as food from an individual to another, but not by force, for people to access useful resources or transfer useful resources to others.

This chapter examined several hypotheses for comparative analysis of food sharing in hunter–gatherer societies using the cases of three groups. The difficulty in generalization through simple comparison can be clearly recognized because of significant differences among hunter–gatherer groups in the quality and quantity of ethnographic information about food sharing. Future studies must undertake further comparative analyses by adding the cases of the Hadza and San peoples in Africa and Australian Aboriginals to examine the propriety and generalizability of the provisional conclusions presented in this chapter.

Notes

1. Hunter–gatherer groups whose food sharing practices have been studied in detail include the San (Dowling 1968; Lee 1979, 1993; Imamura 1993, 1996; Marshall 1961; Maruyama 2016) and the Hadza (Woodburn 1998) in Africa, Australian Aboriginal (Altman and Peterson 1988; Macdonald 2000; Peterson 1993), and the Iñupiat in Alaska (Bodenhorn 2000; Burch 1988; Kishigami 2012) in addition to those described in this chapter. Comparative studies of food sharing in these groups must be undertaken in the future.
2. The definition of sharing is provided by Belk (2010). Sharing is not gift-giving. The first treatise directly addressing "sharing" was apparently that of Price (1975).
3. Kaplan and Hill (1985a) examined theories of kin selection, reciprocal altruism, tolerated theft, and cooperative labor in the case of the Aché and demonstrated that the food sharing of the Aché was not explainable by the respective theories. Gurven, Hill, and Kaplan (2002) attempted to verify the social signal theory, but this theory alone can only inadequately explain the food sharing of the Aché.
4. Kishigami (2016) outlined and examined the gift theory of Marcel Mauss. I regard "food sharing" in hunting and gathering societies as a "total social phenomenon," a concept proposed by M. Mauss (Kishigami 2016).
5. Netsilik Inuit people once had a seal meat partnership system. Under this system, a hunter would give each part of a seal he hunted to a partner having the name of the part. In this case, the parts given to specific people had been decided based on the rule (Van de Velde 1956; Balikci 1970; Damas 1972a, 1972b).
6. In the first sharing in the bowhead whale hunt of the Iñupiat in Alaska, there are rules for distributing specific parts of the whale to the group successfully hunted the whale and to the other groups that had assisted whaling and butchering (Kishigami 2012). Beluga whale hunting in the Nunavik region had related rules for first sharing (Graburn 1969).

7. Oda (2019: 13, 14) pointed out that the food sharing of hunter–gatherers was governed by a system of obscuring the sense of indebtedness likely to be held by the receiver by separating hunters and meat owners and transferring meat through multiple givers and receivers. However, it is difficult for us to verify this remark. I argue that the sense of indebtedness is obscured as a result of socialization. The problem with Oda is to conceive the transfer of all goods together with the concept of exchange and categorize the distribution of them in a single category. The reason is that sharing is not only exchange; it is also a different form of distribution such as giving and redistribution.
8. Peter Freuchen, who studied the Inuit in Greenland, introduced a story of an elderly person demonstrating that the meat sharing of the Inuit was not gifting. "You must not say thank you for your own meat. It is your right to be given that meat. In this land, no one wishes to be dependent on someone else. For that reason, no one gives and receives a gift. If you did, you would be dependent on others. You would create a slave by giving a gift" (Freuchen 1981: 109).
9. Russell Belk defined sharing as "an act or process of sharing our belongings for other people to use them, and/or an act or process of receiving or obtaining goods from others for us to use them" (Belk 2010: 717).

Chapter 5
Food Sharing: Theoretical Considerations

In this book, after reviewing literature on hunter–gatherer food sharing, I introduced food sharing among Canadian Inuit and Alaskan Iñupiat. Additionally, we compared food sharing among the African Aka, South American Aché, and Canadian Inuit. This chapter organizes and examines the terms, features, types, functions, and effects of hunter–gatherer food sharing in order to delineate its significance in human societies, particularly in relation to Arctic research.

It is worth noting here that food sharing is not limited to hunter–gatherers, as small agricultural communities also often engage in the practice (Sillanders 2019; Terashima ed. 2021). Additionally, food is not the only thing shared, but time, places, information, tools, and childcare are also involved; however, this chapter will focus specifically on hunter–gatherer food sharing.

5.1 What is Food Sharing in Hunter–Gatherer Societies like?

Food sharing is a mode of allocation and distribution of foodstuffs. In the case of hunter–gatherers, it is common for individuals or families to give and receive food without any monetary compensation in hunting camps and in settlements. We have defined this act as food sharing, despite the fact that meat is often bought and sold using money in many modern societies. As such, it is more difficult to define food sharing than imagined.

Prior studies have emphasized that hunter–gatherer food sharing occurs within close social relations and distances (Gibson 1985; Ingold 1988; Imamura 1993, 1996; Price 1975; Widlok 2017; Sillander 2019). In other words, the distribution of food tends to occur among those who share in each other's daily lives and activities. Ingold and K. Imamura pointed to food sharing as an aspect of hunter–gatherers' lives and activities, while Widlok (2017) focused on the importance of co-presence. According to Kishigami (2008b, 2013a), food sharing is a part of a subsistence system that is

N. Kishigami, *Food Sharing in Human Societies*, Trust 4,
https://doi.org/10.1007/978-981-16-7810-3_5

inseparable from social organizations, worldviews, values, and ritual practices, which consist of a series of activities, such as hunter–gatherer game spotting, tracking, hunting, butchering, sharing, consumption, and disposal. While these perspectives emerged from surveying local communities, it has now been proposed that food sharing is not dependent upon co-presence and physical proximity. That is to say, Inuit and Iñupiat in various villages, towns and cities of North America are no longer limited to face-to-face contact, and instead use telephones and social media to engage in long-distance food sharing as a result of globalization (Dunn and Gross 2016; Kishigami 2008, 2013c; Stephen R. Braund and Associates 2018).

Given this situation, we ought to reconsider how food sharing is defined. As discussed in Chap. 4, Widlok (2017) proposed a new definition that focuses on recipients and co-presence; although this is interesting, it has its limits. As such, I propose an alternative definition based on Belk's work (2010: 717), which considers both the giver and receiver: food sharing is a social practice for individuals to transfer food to others without force in order to obtain or/and let others to be accessible to food.

Differences between exchange or giving gifts and sharing should be considered in this process (Belk 2010: 717,721; Widlok 2017: 6). First, the former involves a special occasion, whereas the latter is a part of daily activities. Second, in gift-giving, a gift is both received and given in return, but there is usually no gift given back in sharing. Third, the former also involves a show of generosity and authority, while the latter does not and should not. Fourth, "thanks" is usually given for the giving of gifts, although this is not the case in sharing.

5.2 Hunter–Gatherer Sharing Form and Features

What forms or types of food sharing should be considered? As shown in Chap. 1, most anthropological studies on hunter–gatherer food sharing include giving and/or receiving food within a group, such as a hunting parties, camps, or settlements. Based on this, food transfer can be one of three things: unilateral giving, exchange, and redistribution. In addition, they can be further classified based on whether they are done voluntarily, on demand, or based on existing rules (Kishigami 2003b, 2004a). Although Widlok (2017) tried to limit the unilateral provision of goods, services, and knowledge to unilateral giving, we consider sharing to consist of nine types of object transferals based on existing research (see Chap. 1).

We will also examine the characteristics of food sharing in hunter–gatherer societies that are present in the Arctic. As discussed in Chap. 4, there are some commonalities that exist in food sharing across different hunter–gatherer societies. In recent years, Widlok has pointed to co-presence, relatedness, and conversation as components of sharing (Widlok 2017: XVII, 63). Additionally, Bird-David (2019) discussed how there is an underlying link between worldview or environmental views and sharing in hunter–gatherer societies. Further, emphasis is placed on population size and kinship as conditional factors of sharing between hunter–gatherers (Bird-David

2019). Moreover, it has been pointed out that sharing does not involve reciprocity or reciprocal exchange (Widlok 2017:2–17; Woodburn 1998), and distribution is not related to any particular debt obligation (Woodburn 1998). It has also been stated that hunter–gatherer food sharing relates to the size and quantity of game animals. These characteristics will be discussed herein.

5.2.1 Food Sharing and Co-Presence

As introduced in Chap. 4, in places where hunter–gatherers butcher game or in nomadic campsites, food sharing among families and other people that are present, regardless of kinship. The people who share these spaces also share activities and their lives together and are close to each other. Physical proximity and visibility are thus crucial components of food sharing.

However, about 30–50% of Alaska's indigenous people have left their homelands and moved to Anchorage and Seattle, and about 30% of Inuit from northern Canada have migrated to southern places including Ottawa, Edmonton, and Montreal. Entering the twenty-first century, cellphones, e-mail, Facebook, Line, and others are used to keep in touch with family and relatives that are living far away. As such, bowhead whale and beluga whale *maktak* (skin parts with blubber), as well as caribou meat, are sent from settlements in the Arctic regions to families and relatives in southern cities (Dunn and Gross 2016; Kishigami 2008a; Stephan R. Braund and Associates 2018: 26). In many cases, those who wish to eat food from their homelands ask their parents, uncles, aunts, siblings, cousins, or friends that are still living there to send it to them. Food sharing in these cases requires an intimate social relationship, rather than physical proximity or a shared space.

5.2.2 Food Sharing and Close Intimate Relationships

The basis of food sharing is the giving, receiving, and distributing of hunted game or food between certain individuals. In such cases, the relationship between these humans is typically intimate.

Traditionally, hunter–gatherers would not have necessarily paid attention to this kind of relationship—that of kinship—because they shared a common space (Woodburn 1982, 1998). However, in many hunter–gatherer societies today, kinship is a metaphor for intimate relationships.

Bird-David (2019) and Sillander (2019) cautioned against considering kinship essential in food sharing. Yet, it has been pointed out that among Arctic hunter–gatherers, such as Canadian Inuit and Alaskan Iñupiat, as well as the Dolgan and Nganasan of Siberia, extended family relationships pertaining to parents and children or siblings play an important role in daily food sharing (for example, Bodenhorn 2000: 42, 48; Burch 1970, 1975, 1988a: 108–109; Damas 1963, 1972a; Dorais 1997: 1995;

Kishigami 1995, 2000, 2004d, 2007; Nuttall 1991, 1992; Ready 2016; Ziker 2007; Ziker and Fulk 2018; Ziker and Schnegg 2005; Wenzel 1981, 1991, 1995, 2000).

Among Inuit societies in Canada, food sharing is conducted in small campsites regardless of kinship, but in settlements where multiple extended families cohabitate, most food sharing is done within the extended family (Kishigami 1995, 2000, 2007; Wenzel 1995, 2000). In the future, it is hoped that we can similarly elucidate the relationship between food sharing and kinship among hunter–gatherers in Africa, South America, and Asia.

5.2.3 Food Sharing and Conversation

For a long time, hunter–gatherer food sharing has been understood as being associated with altruism and generosity (Widlok 2017: XVII). This perspective has given rise to an image of hunter–gatherers voluntarily giving food to each other (Service 1966). However, it has been established that this image differs from reality.

In his study of Australian Aboriginal food sharing, Peterson (1993) showed that food sharing was dictated by the needs or demands of the receiver, and this demand sharing is found in many hunter–gatherer societies. Widlok (2017) focused on the types of conversations that trigger the start of food sharing, and based on that, pointed out that distribution is started in different ways by the receiver (i.e. through verbal demands, indirect demands, and one's appearance).

Based on Peterson and Widlok's points, it is evident that there are multiple kinds of food sharing in hunter–gatherer societies around the world, and I would like to emphasize that demand sharing is one such type, as shown in Chap. 1. In other words, there are voluntary types of sharing and rule-based types of sharing.

5.2.4 Food Sharing and Worldviews/Environmental Views

As discussed in Chap. 1, the Nayaka in India saw forests as parents that give food to humans and thought that humans should share food like brothers and sisters. The giving environment of the Nayaka provides the ideological basis for food sharing among its people. Additionally, as discussed in Chap. 3, the Alaskan Iñupiat believe that bowhead whales give their lives themselves for people (the whaling captain and his wife) who behave properly towards the whales according to the whales' view. Therefore, whale meat was not to be monopolized by any one individual and was intended to be shared with others (Bodenhorn 2000: 50). They also believe that whales will come to be taken again by those people who share their meat and other edible parts with other people. Therefore, the concept of this relationship between humans and whales forms an ideology that encourages people to share their food. To put it another way, various studies suggest that hunter–gatherer food sharing is intertwined with their worldview and relationship to game.

Thus far, worldviews and environmental views that are related to, or incumbent upon, food sharing have not been studied in depth. Further investigation of the relationship between food sharing and worldviews/environmental views in hunter–gatherer societies in Africa and South America is an important future line of research.

5.2.5 *Food Sharing and Reciprocity*

Since Mauss' anthropological theory of gift (1924), the movement of food and other goods in hunter–gatherer and small agricultural societies has been understood in terms of gift exchange. Gift-giving supposedly occurs voluntarily with the object being given intentionally and unilaterally to the receiver; however, there is an obligation to give, an obligation to receive, and an obligation to return. Basically, gift-giving is a form of exchange. He explains that an object is reciprocated to the first owner (giver) because the spiritual power of the first owner is inherent in the object with a desire for it to be done.

Mauss' theory of gift exchange, along with Sahlins' (1965, 1972) reciprocity model of exchange, have been major influences on food sharing studies for over half a century. Sahlins noted there was an association between social relations (distances) and reciprocity, and proposed a reciprocity model based on transactions, whereby generalized reciprocity occurred between close relations such as relatives, balanced reciprocity occurred between non-intimate relations such as non-relatives, and negative reciprocity occurred between people who were enemies or like enemies.

However, in hunter–gatherer studies, there were cases where food sharing did not occur in a reciprocal exchange. In studying the indigenous people of the Arctic, Pryor and Graburn quantitatively showed that Inuit food sharing was non-reciprocal; similarly, Damas criticized Sahlins' model in his own work (Damas 1972a; Pryor and Graburn 1980). In studying the African Hadza hunter–gatherers, Woodburn (1998) also criticized the reciprocity model and claimed that food sharing was not an exchange, but a redistribution. Furthermore, in investigating the African ≠ Akhoe Hai//om hunter–gatherers, Widlok argued that hunter–gatherer food sharing was a social practice conceptually different from market or gift exchange (Widlok 2017: 1–6; 27).

A recent study on Inuit found that there was a tendency to give and receive between people in families with high earnings and hunting productivity, and the transfer of food between these families has been said to resemble reciprocity or exchange (for example, Ready 2016: 113, 118; 2018). This has also been shown to occur within extended families. In the case of the former, when there is no kinship between the families or households, there is balanced reciprocity, which can be understood as a form of delayed exchange. However, this type of food sharing only accounts for a part of the sharing among Inuit. In the latter cases, the giving results from the norm that families and relatives are meant to help each other, and there is no rule that deems an exchange necessary. Inuit give and receive food to and from family and relatives, but each instance of giving and receiving is an independent action that results in food being transferred between the households of extended family (i.e. there is no obligation or rule to reciprocate this action to a specific person). Therefore, I would like to emphasize that not all Inuit food sharing ought to be considered a gift exchange or reciprocal exchange.

5.2.6 *Food Sharing, Indebtedness, and Superior-Inferior Relationship*

Mauss (1924) pointed out that gift-giving places a burden on the side of the receiver and creates a hierarchical relationship between giver and receiver. In this respect, hunter–gatherers in Africa and the Arctic believe that generosity and authority should not be shown. They think one should not be stingy with food or express thanks to a giver, and in so doing, they make a cultural effort to erase the path of objects being shared (for example, Freuchen 1981: 109; Lee 1993; Marshall 1961; Widlok 2017: 6–7). In particular, Woodburn (1982) argued that the receiver incurs no debt in hunter–gatherer food sharing, meaning that no hierarchical relationship is created between giver and receiver.

In Chap. 4, the point that hunter–gatherer food sharing does not generally create debt or hierarchies was emphasized, and the idea that food should be shared is internalized as a social norm from childhood (Hewlett 2013: Chap. 2; Sillander 2019: 76). Moreover, food sharing in contemporary Inuit society has become a means of obtaining rare food from their homeland, and recipients often express gratitude to those giving the food. We should note that in today's globalized hunter–gatherer societies, food sharing can create indebtedness and superior–inferior relationships.

5.2.7 *Food Sharing, Amounts of Games, and Population Size*

As analyzed in Chap. 4, if game animals are large or there is a large quantity of them, they are more likely to be shared with other people and households; on the other hand, if there is a small amount of game, they tend to not be shared. Furthermore, in nomadic camps of thirty or fewer people, relatively frequent and extensive food sharing is done, but it was also discovered that food sharing among relatives is more common in populated residential settlements with multiple families.

Kelley (1995: 211) estimated that hunter–gatherer band groups average 28.4 people. Additionally, Winterhalder (1986a, 1986b) pointed out that food sharing is most effective in hunting bands with seven or eight hunters. Smith (1988) also points out that food sharing does not operate as effectively when population increases, leading to less face-to-face interactions where people can visibly monitor one another's behavior, and people start storing food. Myers (1988: 58–59) reported that food sharing was no longer done among the growing population of the Australian Aboriginal Pintupi settlement. From these studies, it can be inferred that food sharing works most effectively among populations of less than thirty people.

Based on aforementioned research, the results of food sharing are thought to vary greatly depending on the amount of food available for sharing and the population scale of the group in which sharing is conducted.

5.3 Various Functions and Effects of Food Sharing Among Hunter–Gatherers

Recent food sharing research has transitioned from focusing on theories of origin to theories of function. Some researchers are also of the opinion that one should focus not only on the outcome, but on the process (Widlok 2017: 43; 2019: 27). In this section, we will examine the functions and effects of food sharing. As presented in Chap. 1, these functions include environmental adaptability, economics, politics, socialization, culture, and psychology.

5.3.1 Environmental Adaptation

Food sharing changes based on the season. Sharing, mobility, and storage of food are potential ways in which humans can maintain a stable food supply in an unstable environment (Widlok 2017: 46). Hunter–gatherer food sharing acts as risk avoidance tactic for food acquisition, and the theory is that it improves adaptability to live in a natural environment. In many hunter–gatherer societies, because hunting can be an unstable means of getting food, it is thought that a successful hunter would give some of their game to others who were unsuccessful to improve their likelihood of survival. Therefore, this approach serves to reduce risk as a kind of insurance (Widlok 2017: 55).

This theory is the most popular in explaining the origins of food sharing and its practice among hunter–gatherers. However, in many modern hunter–gatherer societies where the food supply is stable, food sharing is not essential for survival and adapting to the environment. In Australian Aboriginal societies, food sharing has been reported to have other functions, such as affirmation and coordination of social relations among specific people (Macdonald 2000; Peterson 1993).

5.3.2 Economic Function

In terms of economic function, food sharing among hunter–gatherers has a leveling function that prevents certain individuals or families from monopolizing the accumulation and consumption of food within a group (Woodburn 1981). As discussed in Chap. 4, food is entirely distributed within the small camps of the Aka hunter–gatherers, and each person and family receive virtually the same quality and quantity of food. In these cases, food sharing can be said to eliminate economic disparities between individuals and families. However, in settlements where many individuals and families are present, unless large game like bowhead whales, beluga whales, or elephants are hunted and distributed to all members, game and food tend to be given to relatives and those living in close proximity, which cannot be said to eliminate economic disparity throughout the settlement.

In modern Inuit settlements in Nunavik, Canada, attempts have been made to use the hunter support program to provide equal food to everyone, as mentioned in Chap. 2. However, routine food sharing tends to occur within each extended family, and there are limits to the extent that economic disparities can be reduced across settlements (Kishigami 2000, 2018). In other words, while this has the function of reducing economic disparity within an extended family, it does not fully remove inequality between extended families. Ready's (2016) point that contemporary Inuit food sharing does not contribute to the well-being of an entire settlement is accurate in this context.

5.3.3 Political Function

Food sharing is also related to preventing certain individuals or families from monopolizing goods, and it is theorized that this promotes political autonomy and equality between individuals or/and families. While this hypothesis holds for small hunter–gatherer populations that migrate regularly, such as the Aka, it is not valid for more stratified hunter–gatherer societies, such as the Northwest Coast People of North America.

For the Iñupiat on the northwestern coast of Alaska, who form a whaling population and catch large bowhead whales, and the Haida on the northwestern coast of North America, where fishery resources are abundant, meals and food are distributed to participants at nalukataq feasts and potlatch ceremonies, respectively. In each of these societies, there is an authority figure that wields considerable political and economic power (in the former case, a whaling captain with an umiaq (a large whaling boat); and in the latter case, the chief of an extended family group). Hence, the act of sharing does not necessarily create political egalitarianism. Rather, they gain more political power and social influence through food distribution.

In addition, even in today's hunter–gatherer societies, those who provide frequently and in greater abundance are shown to gain better social standing and prestige (Collings et al. 2016: 36; Kishigami 2018; Widlok 2017: 50). Moreover, it has been reported that several politically ambitious Inuit use food sharing to gain political influence and status (Ready 2016: 141–142).

Based on this, it can be said that hunter–gatherer food sharing does not necessarily result in political egalitarianism.

5.3.4 Social Function

In its most simple form, food sharing is the act of someone with food giving another person a portion of that food. This act is often performed between two people with an intimate social relationship, such as a parent and a child or siblings, and the transfer of food between the two makes visible and confirms that relationship. In this way, hunter–gatherer food sharing is closely linked to certain social relationships and fulfills the function of maintaining, changing, and creating such relationships.

Researchers including Sillander (2019) and Bird-David (2019) have worked to clarify the close link between food sharing and social relationships. For example, in modern Arctic indigenous societies, food sharing is known to contribute to the confirmation and maintenance of extended family relations (Kishigami 2000), as well as the social integration of communities (Chabot 2003; Collings, Wenzel, and Condon 1998).

5.3.5 Cultural Function

Food sharing reflects the worldviews, relationship to animals, norms, and values of the people who engage in it. These things are also symbolized and reproduced through food sharing. While the Nayaka's concept of the giving environment (forests) and the Iñupiat's strong ideas about the mutually beneficial relationship between bowhead whales and people prompt sharing practices, each society's worldview/environmental view is reproduced through the practice of food sharing (Bird-David 1990; Bodenhorn 2000). In particular, food sharing among the indigenous peoples of the Arctic plays a role in tying them to the environment and social beings other than human beings (Fienup-Riordan 1999; Laugrand and Oosten 2014; Stairs and Wenzel 1992).

In addition, it can be said that the practice of food sharing generates, maintains, and strengthens the identities and ethnicities of people by openly saying "this is our way" and "this is our way of life" (for example, Condon et al. 1995; Bodenhorn 2000: 47).

5.3.6 *Psychological Function*

In the field of cultural anthropology, social factors that lead to food sharing have often been explained using cultural factors and factors related to ecological adaptation, but some studies have also explored the psychological states of people. In particular, it is known that people often say that one of the reasons they give food is because they derive pleasure or joy from doing so[1] (Bodenhorn 2000: 27; Fienup-Riordan 1983; Lewis 2019).

For example, Iñupiat whaling captains have often said to me that they engage in whaling to provide products to villagers because the villagers need and want whale meat. Similarly, when surveying Canadian Inuit, Wenzel found that Inuit hunters hunted for the well-being of the settlement as a whole (Wenzel 1991). Subsistence activity including food sharing is an aim itself for many Arctic people (Fienup-Riordan 1983).

Based on my experiences in Canada and Alaska, a good hunter provides a lot of game to relatives and other members of the settlement and has a high level of fame and a prestigious reputation. Furthermore, they seemed to derive satisfaction from being admired by others for these actions. In relation to this, it has been shown that empathy and joy play strong roles in food sharing among apes (Matsuzawa 2018: 202–203). Therefore, further examination of the relationship between food sharing and emotions is an important topic for future research.

As we have discussed thus far, hunter–gatherer food sharing is a social practice with many functions and effects. Therefore, to better understand it, it is essential that we try to comprehensively detail what kind of food sharing occurs at a given time and place.

5.4 Sharing as a Human Characteristic

While the sharing of food between many animals and their children may be an original form of food sharing, we seldom see food sharing between individual adult animals. According to Jaeggi and Gurven (2013: 188), of the 69 primate species, 30 don't share food at all, 39 only share food with their own offspring, and 15 also share food with other individual adults. In primates other than humans, tolerated theft is a common way of sharing, and there is almost no exchange seen[2] (Konma 1999: 12; Matsuzawa 2018: 125; Widlok 2017: 32; Yamagiwa: 1997: 51). That is to say, extensive food sharing is characteristic of human beings because of their excellent memory and thinking faculties, which have come about as a consequence of their language ability through human evolution. Although humans continue to evolve and may change further, I believe that greater understanding of food sharing can help us better understand and elucidate the nature of modern human beings.

Human beings are, of course, living things, and food sharing is conducted based on their biology; however, the major difference between humans and other animals in this regard is that humans have developed food sharing as a social and cultural practice during their evolution. The characteristic of human food sharing is uniform, in that it is used for access to food, although various forms have existed at different times in different places. Therefore, human food sharing that occurs at a particular time and place cannot be understood if we ignore its social, cultural, and historical contexts.

In this book, we considered the cases of Inuit and Iñupiat food sharing. In these cases, it is clearly a social practice with multiple functions and effects, and cannot be reduced to one source or factor. Moreover, this social practice has changed over time.

A most interesting point is that as hunter–gatherer societies around the world have undergone political integration into nations (as a result of economic globalization), they continue to carry out food sharing as a central social practice, even as their lifestyles have changed. It can also be said that food sharing, with its complex functions and effects tied so strongly into their lives, is equivalent to what Mauss termed a "total social fact" (fait social total) within a small scale community. Even today, it continues to play an important role in producing and maintaining their micro–social order.

The act of sharing occurs not only with food, but exists in almost all human societies in terms of the sharing and shared use of child-rearing, knowledge, information, spaces, and tools. Thus, as Widlok (2017) demonstrated, the importance of sharing practices is beginning to be noticed even in modern non-hunter–gatherer societies. While globalization has led to division and the individualization of people around the world, we are also realizing how connected people are to each other. When thinking about the future of human societies, I believe that the act of sharing, which has impacted the micro-dimensions, will become more and more crucial in the formation and maintenance of social relationships.

5.5 Conclusion

Over a long period of evolution, humans have acquired a great proficiency for language, and this has also led to advanced thinking, better memories, and more extensive communication. Against this background, hunter–gatherers began the social practice of food sharing, which continues to this day. The methods and objects involved have differed depending on the time, region, and group, but the fact that they persist even now suggests that they are an effective means of adapting to the ever-changing natural and social environments.

Compared to other animals like primates, the trait of sharing goods and services such as food can be said to be specific to humans. In particular, it is known that only humans make reciprocal exchanges, but I think the biggest difference between humans and other living things is the giving or letting go of things, especially food, to allow others to have them. Giving and letting go is the foundation for reciprocal exchange, and by extension, the formation of social relationships. Even as times change, this is something that I think of as one of the basic tenants of human behavior.

Although originally practiced in close, face-to-face social relationships, food sharing is beginning to be done in different ways in our globalized society. Although we have focused on micro cases involving food in this book, the sharing economy takes on many forms in today's globalized society, such as ride sharing, room sharing, blood donation, food programs, and universal basic income, all of which are attracting great attention (Kishigami 2016; Widlok 2017). This human behavior has grown from individuals, families, and households to various groups, such as companies and NGOs, nations, and international organizations, such as the United Nations. In many contexts, their function is to try and reduce economic disparities and provide disaster relief[3]. On the other hand, it should be remembered that depending on the situation and means, hunter–gatherer food sharing might not lead to relief or improvement of people's lives, but may maintain and reproduce the existing political and economic systems, which favor people that already have political and economic power.

Food sharing is a combination of the wisdom and practices of humankind. I argue that it is important to continue these practices, while also critically examining their fate from an anthropological perspective.

Notes

1 Matsuzawa, who studied chimpanzees, said "Sharing reciprocally constitutes a joy in itself" (Matsuzawa 2018: 202).

2 Yamagiwa (1997: 51) characterized sharing among apes as non-reciprocal and sharing among humans as reciprocal. In the former case, the gift-giving is one-sided and involves begging, whereas the latter case involves giving and exchange (giving to one another).

3 Analysis of economic support, disaster relief, blood donation, etc. in modern society is mainly conducted from the perspective of Mauss' gift-giving theory, although the theory of sharing can also be used for this purpose (Kishigami 2016; Widlok 2017).

Brief CV of Nobuhiro Kishigami

(Date: Dec 1, 2021)

(Name) KISHIGAMI, Nobuhiro

(Sex) male

(Birth) 04/04/58

(Nationality) Japanese

(Current Position, Affiliation and Contact Address)

Professor

Center for Cultural Resource Studies

National Museum of Ethnology (Minpaku)

10-1 Senri Expo Park

Suita City, Osaka, Japan 565-8511

(Phone) 81-6-6878-8255 (office)

(e-mail) inuit@minpaku.ac.jp

(Education)

B.A. (Sociology) Waseda University, Tokyo, Japan 1981(March)

M.A. (Sociology) Waseda University, Tokyo, Japan 1983(September)

Ph.D. Candidate (Anthropology) McGill University, Montreal, Canada 1986 (enrolled 1983–1989)

Ph.D. (Anthropology) Graduate University for Advanced Studies (Sokendai), Kanagawa, Japan 2006 (March)

(Primary Job Career)

1989.4—1990.9 Research Fellow (Assistant Professor), Waseda University, Tokyo, Japan

1990.10—1992.3 Fulltime Lecturer, Hokkaido University of Education, Hakodate, Japan

1992.4—1995.12 Associate Professor, Hokkaido University of Education, Hakodate, Japan

1996.1—2005.10 Associate Professor, National Museum of Ethnology, Osaka, Japan

N. Kishigami, *Food Sharing in Human Societies*, Trust 4,
https://doi.org/10.1007/978-981-16-7810-3

1997.1—2005.10 Associate Professor, The Graduate University for Advanced Studies, Japan

2005.10—up to now, Professor, National Museum of Ethnology, Osaka, Japan

2005.10—2018.3 Professor, The Graduate University for Advanced Studies, Osaka, Japan

2009.4—2012.3 Director, Department of Advanced Studies in Anthropology, National Museum of Ethnology, Osaka, Japan

2012.4—2013.3 Director, Center for Research Development, National Museum of Ethnology, Osaka, Japan

2013.4—2016.3 Deputy Director General, National Museum of Ethnology, Osaka, Japan

2017.4—2018.3 Director, Center for Cultural Resource Studies, National Museum of Ethnology, Osaka, Japan

2018.4—2022.3 Executive Director, National Institutes for the Humanities, Tokyo, Japan

(Membership) Japanese Society of Cultural Anthropology (JSCA), Japanese Association of Canadian Studies (JACS), International Association for Arctic Social Sciences (IAASS)

(Committee and Board Members)

2018 April—Up to now: Council Member, Japanese Society of Cultural Anthropology

2016—Up to Now Associate Editor, Journal of Anthropological Research

2012 April—Up to now: Board Member, Japanese Association of Canadian Studies

2012 April—Up to now: Board Member, Japanese Association of Ethnic Art Studies

2012 April—2016 March: Editorial Board Member, Japanese Review of Cultural Anthropology

2011 October—2017 September: Member (Area Studies), Science Council of Japan (SCJ)

2008 April—2008 March and 2012 April—2016 March: Board Member, Japanese Society of Cultural Anthropology

2006 April—2008 March: Council Member, Japanese Society of Cultural Anthropology

(Major Research Grant)

April, 2019—March, 2024 Grant-in-aid for Scientific Research (Japan Society for the Promotion of Science): A Comparative Study of Emergence, Current Status and Future of Indigenous Cultures in the Alaska and Northwest Coast Regions of North America. 33.5 million yen (About US$320,000)

April, 2015—March 2019 Grant-in-aid for Scientific Research (Japan Society for the Promotion of Science): An Anthropological Study of Whaling Cultures in the Globalizing World. 26 million yen (About US$220,000)

April, 2009—March, 2014 Grant-in-aid for Scientific Research (Japan Society for the Promotion of Science): Aboriginal Subsistence Whaling and Indigenous Rights in North America. 14.82 million yen (About US$115,000)

April, 2003—March 2007 Grant-in-aid for Scientific Research (Japan Society for the Promotion of Science): Indigenous Distribution and Management of Marine Resources. 44.46 million yen (About US$400,000)

April, 1999—March, 2002 Grant-in-aid for Scientific Research (Japan Society for the Promotion of Science): Indigenous Use and Management of Marine Resources. 18.35 million yen (About US$ 150,000)

(Research Topic)

Social change among the Nunavik Inuit, Urban Inuit in Montreal, Indigenous Use and Management of Marine Resources, aboriginal subsistence whaling (Inupiat in Alaska), Development and Indigenous People, etc. in Cultural Anthropology.

English Publication List (Nobuhiro KISHIGAMI)

(Edited Books in English)

2021 *World Whaling: Historical and Contemporary Studies* (SES No. 104) Osaka: National Museum of Ethnology.

2013 *Anthropological Studies of Whaling* (SES No.84) Osaka: National Museum of Ethnology. Edited with Hisashi Hamaguchi and James M. Savelle

2005 *Indigenous Use and Management of Marine Resources* (SES No.67) Osaka: National Museum of Ethnology. Edited with James M. Savelle

2000 *The Social Economy of Sharing: Resource Allocation and Modern Hunter-Gatherers* (SES No.53) Osaka: National Museum of Ethnology. Edited with George W. Wenzel and Grete Hovelsrud-Broda.

(Journal articles and book chapters in English)

2021 World Whaling and Recent Research Trends. in N. Kishigami (ed.) *World Whaling: Historical and Contemporary Studies* (Senri Ethnological Studies No. 104), pp. 1–30. Osaka: National Museum of Ethnology.

2021 Food Security, Food Sovereignty, and Bowhead Whale Hunts among the Iñupiat in Utqiaġvik, Alaska, USA. in N. Kishigami (ed.) *World Whaling: Historical and Contemporary Studies* (Senri Ethnological Studies No. 104), pp. 93–112. Osaka: National Museum of Ethnology.

2016 An Info-Forum Museum for Cultural Resources of the World: A New Development at the National Museum of Ethnology. In Ito, Atsunori (ed.) *Re-Collection and Sharing Traditional Knowledge, Memories, Information, and Images: Challenges and the Prospects on Creating Collaborative Catalog* (Senri Ethnological Reports 137), pp. 25–33. Osaka: National Museum of Ethnology.

2016 Revival of Inuit Bowhead Hunts in Arctic Canada. *Japanese Review of Cultural Anthropology* 16:43–58.

2015 Low-income and Homeless Inuit in Montreal, Canada: Report of a 2012 Research. *Bulletin of the National Museum of Ethnology* 39(4): 575–624.

2013 Anthropological Research on Whaling: Prehistoric, Historic and Current Contexts. (written with James M. Savelle), in N. Kishigami, H. Hamaguchi, and J. M. Savelle (eds.) *Anthropological Studies of Whaling* (Senri Ethnological Studies 84), pp. 1–48. Osaka: National Museum of Ethnology.

2013 Aboriginal Subsistence Whaling in Barrow, Alaska. In N. Kishigami, H. Hamaguchi, and J. M. Savelle (eds.) *Anthropological Studies of Whaling* (Senri Ethnological Studies 84), pp. 101–120. Osaka: National Museum of Ethnology.

2013 The Inuit's Migration Patterns and Drastic Population Increase in Urban Centers of Canada. In Klaus-Dieter Ertler and Patrick Imbert (eds.) Cultural Challenges of Migration in Canada (Canadiana 12), pp. 65–73. Frankfurt am Main: Peter Lang.

2013 What Is a Subsistence Activity?: With a Special Focus on Beluga Whale Hunt by Inuit in Arctic Canada. *Jinbun-Ronkyu* (Journal of the Society for Liberal Arts at Hokkaido University of Education, Hakodate, Japan) 82:79–90.

2013 On Sharing of Bowhead Whale Meat and Maktak in an Inupiat Community of Barrow, Alaska, USA. *Bulletin of the Hokkaido Museum of Northern Peoples* 22: 1–19.

2010 Climate Change, Oil and Gas Development, and Inupiat Whaling in Northwest Alaska. *Etudes/Inuit/Studies* 34 (1): 91–107.

2008 Homeless Inuit in Montreal. *Etudes/Inuit/Studies* 32 (1): 73–90.

2008 Trends in Japanese Cultural Anthropological Research on Whaling Cultures. *Japanese Review of Cultural Anthropology* 9: 71–99.

2008 Notes on Information and Food-resource Flow among the Urban Inuit in Montreal, Canada. *Jinbun-Ronkyu* (Journal of the Society for Liberal Arts at Hokkaido University of Education, Hakodate, Japan) 77: 235–246.

2007 Indigenous Trade and Social Change of the Siberian Yupik Eskimos in the Bering Strait Region during the 18–20th Centuries. *Jinbun-Ronkyu* (Journal of the Society for Liberal Arts at Hokkaido University of Education, Hakodate, Japan) 76: 39–57.

2006 Inuit Social Networks in an Urban Setting. In P. Stern and L. Stevenson (eds.) *Critical Inuit Studies: An Anthology of Contemporary Arctic Ethnography*, pp. 206–216. Lincoln, Nebraska: University of Nebraska Press.

2005 Co-Management of Beluga Whales in Nunavik (Arctic Quebec), Canada. In N. Kishigami and J. Savelle (eds.) *Indigenous Use and Management of Marine Resources* (Senri Ethnological Studies 67), pp. 121–144. Osaka: National Museum of Ethnology.

2004 Trends in Native North American Studies in Japan since the 1990s. *Japanese Review of Cultural Anthropology* 5: 91–121.

2004 A New Typology of Food-sharing Practices among Hunter-gatherers, with a Special Focus on Inuit Examples. *Journal of Anthropological Research* 60: 341–358.

2004 Cultural and Ethnic Identities of Inuit in Canada. In T. Irimoto and T. Yamada (eds.) *Circumpolar Ethnicity and Identity* (Senri Ethnological Studies 66), pp. 81–93. Osaka: National Museum of Ethnology.

2002 Urban Inuit in Canada: A Case from Montreal. *Indigenous Affairs* 3–4/2002, pp. 54–59.

2002 Inuit Identities in Montreal, Canada. *Etudes/Inuit/Studies* 26 (1): 183–191.

2002 Living as an Inuk in Montreal: Social Networks and Resource Sharing. *Jinbun-Ronkyu* (Journal of the Society for Liberal Arts at Hokkaido University of Education, Hakodate, Japan) 71: 73–84.

2000 Contemporary Inuit Food Sharing and Hunter Support Program of Nunavik, Canada. In G.W. Wenzel, G. Hovelsrud-Broda and N. Kishigami (eds.) *The Social Economy of Sharing: Resource Allocation and Modern Hunter-Gatherers* (Senri Ethnological Studies 53). pp. 171–192. Osaka: National Museum of Ethnology.

1999 Why Do Inuit Move to Montreal?: A Research Note on Urban Inuit. *Études/Inuit/Studies* 23 (1–2): 221–227.

1999 Life and Problems of Urban Inuit in Montreal: Report of 1997 Research. *Jinbun-Ronkyu* (Journal of the Society for Liberal Arts at Hokkaido University of Education, Hakodate, Japan) 68: 81–109.

1998 Names and Naming of the Reindeer Koryak and Even: Research from Esso District on Kamchatka Peninsula, Russia. In M. Oshima (ed.) *Subsistence, Society and Folk Art of Indigenous Peoples in Kamuchatka Region*, pp. 73–81, Otaru, Japan: Center for Language Studies, Otaru University of Commerce.

1997 Personal Names, Name Souls and Social Change Among Canadian Inuit: A Case Study of Akulivik Inuit, Nunavik, Canada. In T. Yamada and T. Irimoto (eds.) *Circumpolar Animism and Shamanism*, pp. 151–166. Sapporo: Hokkaido University Press.

1995 Extended Family and Food Sharing Practices among the Contemporary Netsilik Inuit: A Case Study of Pelly Bay. *Journal of Hokkaido University of Education* Section I. B 45 (2): 1–9.

1994 Why Become Christians?: Hypotheses on the Christianization of the Canadian Inuit. In T. Irimoto and T. Yamada (eds.) *Circumpolar Religion and Ecology: Anthropology of the North*, pp. 221–235. Tokyo: University of Tokyo Press.

1993 Dogs in the Spiritual World of Traditional Inuit Society of Canada–With Special Reference to Dogs in the Traditional Netsilik Inuit Society. Hokkaido Museum of Northern Peoples (ed.) *Animals in the Spiritual World of Northern Peoples: The 7th International Abashiri Symposium*, pp. 15–26, Abashiri, Japan: Association for Northern Cultural Promotion.

Book Review and Other Publication

2018 Comments on "Why Wage Earners Hunt Food Sharing, Social Structure, and Influence in an Arctic Mixed Economy" *Current Anthropology* 59(1): 89–90.

2017 "History and Current Status of Indigenous People in Canada, International Symposium September 9, 2017" *Minpaku Anthropology Newsletter* 45: 14–15.

2017 "Minpaku Info-forum Museum Project: Progress Report" *Minpaku Anthropology Newsletter* 45: 4–5.

2016 Comments on "Relocation Redux: Labrador Inuit Population Movements and Inequalities in the Land Claims Era" *Current Anthropology* 57(6): 798.

2016 Book review: A Tale of Three Villages: Indigenous-Colonial Interactions in Southwestern Alaska, 1740-1950. Liam Frink. Tucson: University of Arizona Press,

2016, 184PP. $55.00, Cloth. ISBN 978-0-8165-3109-7. *Journal of Anthropological Research* 72(4): 566–567.

2014 "Unknown Land, Greenland: Its Nature and Culture, Thematic Exhibition September 4–November 18, 2014" *MINPAKU Anthropology Newsletter* 39: 14.

2014 Comment on 'The Rotten Renaissance in the Bering Strait'. *Current Anthropology* 55(5): 638–639.

2014 Comparative Studies of Indigenous Cultures around the North Pacific Rim: Focusing on Indigenous Rights and Marine Resource Utilization. *Minpaku Anthropology Newsletter* 38: 10–11.

2013 Preface. (written with J. M. Savelle and H. Hamaguchi) In N. Kishigami, H. Hamaguchi, and J. M. Savelle (eds.) *Anthropological Studies of Whaling* (Senri Ethnological Studies 84), pp. i–ii. Osaka: National Museum of Ethnology.

2013 Concluding Remarks. (written with J. M. Savelle and H. Hamaguchi) In N. Kishigami, H. Hamaguchi, and J. M. Savelle (eds.) *Anthropological Studies of Whaling* (Senri Ethnological Studies 84), pp. 355–357. Osaka: National Museum of Ethnology.

2013 (Research Report) *Sharing and Distribution of Whale Meat and Other Edible Whale Parts by the Inupiat Whalers in Barrow, Alaska, USA*. Osaka: Kishigami's Office, National Museum of Ethnology.

2011 Exhibition: Devotion of the Arts of Living: Daily Life among the Aynu of the Kurile, Sakhalin and Hokkaido Islands–Objects from the Late 19th-Early 20th Century Collections of Ethnological Museums in Germany and Japan (Special Exhibition October 6–December 6, 2011) *MINPAKU Anthropology Newsletter* 33: 12.

2011 Whaling Cultures of the World: Past, Present, and Future. *MINPAKU Anthropology Newsletter* 32: 17.

2011 Comment on James Cliford's Lecture "Repatriation and the Second Life Heritage: Return of the Masks in Kodiak, Alaska" *Bulletin of the National Museum of Ethnology* 35(4): 739–741.

2010 Book Review: *Furs and Frontiers in the Far North: The Contest among Native and Foreign Nations for the Bering StraitFur Trade.*(The Lamar Series in Western History.) by John R. Bockstoce. New Haven: Yale University Press. *The American Historical Review* 115: 814–815.

2009 Voices from the Land, Visions of Life: Beauty Created by the Indigenous Peoples of Canada-Special Exhibition-September 10–December 8, 2009. *MINPAKU Anthropology Newsletter* 29: 10–11.

2009 Book Review: *Soul Hunters: Hunting, Animism, and Personhood among the Siberian Yukaghirs*, by R. Willerslev. *Journal of Anthropological Research* 65: 169–171.

2008 (With M. Lee) Les Inuit urbains. *Etudes/Inuit/Studies* 32 (1): 5–7.

2008 (With M. Lee) Urban Inuit. *Etudes/Inuit/Studies* 32 (1): 9–11.

2007 Book Review: *Social Life in Northwest Alaska: The Structure of Inupiaq Eskimo Nations*, by E. S. Burch, Jr. *Journal of Anthropological Research* 63: 434–436.

2006 Anthropological Research and Inuit Community Development in Montreal, Canada. *Minpaku Anthropology Newsletter* 23: 3–4.

2006 Development Cooperation of Norway: University of Bergen, Chr. Michelsen Institute, and NGOs. *Minpaku Anthropology Newsletter* 23: 14.

2006 Book Review: *The Whales, they Give Themselves: Conversations with Harry Brower, Sr.* (2004). *Journal of Anthropological Research* 62: 128–129.

2005 General Introduction: Marine Resources and Anthropology. In N. Kishigami and J. Savelle (eds.) *Indigenous Use and Management of Marine Resources* (Senri Ethnological Studies 67), pp. 1–9. Osaka: National Museum of Ethnology.

2005 Concluding Remarks. In N. Kishigami and J. Savelle (eds.) *Indigenous Use and Management of Marine Resources* (Senri Ethnological Studies 67), pp. 445–448. Osaka, National Museum of Ethnology.

2004 Role of Social Sciences in Development Projects: International Workshop 6–7, 2004. *Minpaku Anthropology Newsletter* 19: 14.

Major Presentations in English

2017 Indigenous trading networks across the Bering Strait from the eighteenth to the twentieth centuries. Panel 262 at the 10th International Convention of Asian Scholars, Chiang Mai Exhibition and Convention Center, Thailand (July 23, 2017).

2017 Whaling Right and Animal Welfare. Session 8.10 "Recognition, protection and management of indigenous land resource rights in the Arctic—Domestic and international approaches", the 9th International Congress on Arctic Social Sciences (ICASS IX), University of Umea, Sweden (June 11, 2017).

2017 Recent Trends in Inuit Migration to Southern Regions of Canada: A focus on Montreal. Session 22.2b "Arctic Dwellers on the Move: Studying social dynamics beyond the ethnicity lens", the 9th International Congress on Arctic Social Sciences (ICASS IX), University of Umea, Sweden (June 8, 2017).

2016 Sharing of Bowhead whale meat among the Inupiat in Barrow, Alaska, USA. A paper read at the International Conference "SHARING—The Archaeology & Anthropology of Hunter-Gatherers" at McDonald Institute for Archaeological Research, University of Cambridge, UK (2016.9.20).

2015 A Comparative Study of Contemporary Indigenous Whale Hunts in North America. Session 42 "Aboriginal Whaling and Identity in the Twenty-First Century" of CHAGS 11, University of Vienna, Austria (September 9, 2015).

2015 Inuit in Urban Centers: A Case Study from Montreal, PQ, Canada. 2015 Integrated International Conference of The Japan Studies Association of Canada (JSAC), Japanese Association for Canadian Studies (JACS), and Japan-Canada Interdisciplinary Research Network (JCIRN) at Embassy of Canada in Tokyo, May 21, 2015.

2015 The Impact of Climate Change on Aboriginal Subsistence Whaling in Northwest Alaska: Inupiat Whalers, Bowhead Whales and Oil/Gas Development. Session B8: From Human Security to Geopolitical Dynamics in the Global Arctic: The Global Implications of Rapid Environmental, Economic, and Societal Change. The Fourth International Symposium on the Arctic Research (ASSW 2015), Toyama International Conference Center, Toyama, Japan (2015.4.30).

2014 The Inuit Bowhead Hunt and Indigenous Rights in Canada. Paper read at IUAES 2014, Makuhari Messe, Chiba, Japan (May 15, 2014).

2014 Low-income and Homeless Inuit in Montreal. Public Lecture "Tokyo and Montreal—The Urban Lives of the Japanese Ainu and the Canadian Inuit Groups" (organized by the Japan Foundation, Canada), at Thomson House, McGill University (March 14, 2014).

2014 The History and Present Situation of Comparative Studies of Indigenous Cultures around the North Pacific Rim. The International Symposium "Comparative Studies of Indigenous Cultures around the North Pacific Rim: Focusing on Indigenous Rights and Marine Resources", National Museum of Ethnology, Osaka, Japan (January 11, 2014).

2013 Food Security Problems of the Inupiat in Northwest Alaska. International Symposium "Anthropological Study of Food Security" at Osaka University Hall, Osaka University (December 19, 2013).

2013 Homeless Inuit of Urban Centers in Canada: Results from Montreal Research. Paper read at the 73rd Annual Meeting of the Society for Applied Anthropology, Denver Marriott City Center Hotel, Denver, Colorado, USA (March 21, 2013).

2013 "Sharing and Distribution of Bowhead Whale Meat and Maktak in an Inupiat Community of Barrow, Alaska, USA." Public Lecture delivered at UIC Science Center, Barrow, Alaska, USA (February 9, 2013).

2012 The Inuit's Migration Patterns and Drastic Increase in Urban Centers of Canada. ICCS (International Council for Canadian Studies) International Conference, University of Ottawa (May 23,2012).

2011 Aboriginal/Indigenous Subsistence Whaling in Alaska, USA. (paper read by Dr. David Koester) At the session 02.25: Salomon Fishing and Whaling Peoples of the North Pacific, ICASS (International Conference for Arctic Social Sciences) VII, Akureyri, Iceland on June 22, 2011 (2011 June).

2011 Sharing and Distribution of a Bowhead Whale among the Inupiat in Barrow, Alaska. (paper read by Dr. Peter Collings) At the session 03.60: The Social Economy of Sharing in the North American Arctic, ICASS (International Conference for Arctic Social Sciences) VII, Akureyri, Iceland on June 25, 2011 (2011 June).

2011 "The Contemporary Inupiat Whaling in Barrow, Alaska." A Public Lecture Delivered for the Barrow Arctic Science Consortium's Schoolyard Project, 15 January, The UIC Barrow Arctic Research Center, Barrow, Alaska, USA.

2008 "A Cultural Anthropological Study of Subsistence Activities with a Special Focus on Indigenous Hunting and Fishing in the Arctic Regions." The 6th International Congress of Arctic Social Sciences (ICASS VI), 23 August, 2008, Ilimmarfik, Nuuk, Greenland.

2008 "Homeless Inuit in Montreal, Canada." The 6th International Congress of Arctic Social Sciences (ICASS VI), 25 August, 2008, Ilimmarfik, Nuuk, Greenland.

2006 "Homeless Inuit in Montreal: How to Survive in the Hopeless City." The Canadian Anthropology Society Conference "Aboriginal Peoples in Cities, Identity and Human Nature", 16 May, 2006, Concordia University, Montreal, PQ, Canada.

2004 "Information and Material-resource Flow among the Urban Inuit: Research from Montreal, Canada." Revised AAA Session. Peripherals@the center.com: Inuit in Urban Spaces, 17 November, 2004, Gifford Room, Kroeber Hall, University of California, Berkeley.

2004 "Contemporary Inuit Food Sharing." The 5th International Congress of Arctic Social Sciences (ICASS V), 22 May, 2004, University of Alaska, Fairbanks.

2003 "Co-management Problems and the Use of Common Marine Resources." International Association for the Study of Common Property (IASCP), Northern Polar Regional Conference "Joining the Northern Commons: Lessons for the World, Lessons from the World", 19 August, 2003. The Captain Cook Hotel, Anchorage, USA.

2002 "Indigenous Use and Co-management of Beluga Whales in Nunavik (Arctic Quebec), Canada." Paper read at the Japanese Ministry of Education International Symposium 2002, New Interdisciplinary Approaches to the Study of Indigenous Use and Management of Migratory Marine Resources, 3 December, 2002, National Museum of Ethnology, Osaka, Japan.

2002 "Indigenous Trade and Social Change of the Siberian Eskimos in the Bering strait Region during the 18–20th Centuries. The International Symposium "The Raven's Arch: Jesup North Pacific Expedition Revisited", (at Tenjin-yama Guest House, Sapporo City, Japan, 27th, October, 2002).

2002 "A Typology of Food Sharing Practices among Hunter-gatherers, with a Special Focus on Inuit Examples." The 9th International Conference on Hunting and Gathering Societies (CHAGS9), 10 September, 2002, Edinburgh Conference Centre, Heriot-Watt University, Edinburgh, Scotland.

2001 "Living as an Inuk in Montreal: Social Networks and Resource Sharing. Paper read at the 100th AAA Meeting "Boas and the Beyond: the State of the art in Inuit Studies", 28 November, 2001, Washington D.C.

2001 "A New Form of Resource Sharing Practices among the Contemporary Nunavik Inuit in Canada." 4th International Congress of Arctic Social Sciences (ICASS IV), 17th, May, 2001, Quebec City, Canada.

2001 "Inuit Identities in Montreal, Canada." 4th International Congress of Arctic Social Sciences (ICASS IV), 17 May, Quebec City, Canada.

2000 "Identities of Inuit in Canada: Being an Inuk and Being one of the Inuit." International Conference on Ethnicity and Identity in the North, ICNSA 3, 12–14 October, Sapporo, Japan.

2000 "Hunter Support Program of Nunavik, Canada: A Case Study from Akulivik." 12th Inuit Studies Conference, 23–26 August, University of Aberdeen, Scotland.

1998 "Contemporary Inuit Food Sharing and Hunter Support Program of Nunavik, Canada" The 8th International Conference on Hunting and Gathering Societies, 27 October, 1998, National Museum of Ethnology, Osaka, Japan.

1995 "Personal Names, Name Souls and Social Change among Canadian Inuit: A Case Study of Akulivik Inuit, Nunavik, Canada." International Conference on Animism and Shamanism in the North (ICNSA) Session 4 "Animism and Shamanism in the Inuit/Eskimos", 14 October, 1995, Sapporo, Japan.

1994 "Recent Trends and Current State of Inuit and Yupik Studies in the Field of Cultural Anthropology in North America: 1984–1993." The International Symposium and Native Folk Festival: Art of Native Peoples in the North, 25–29, June, 1994, Magadan, Russia.

1994 "Extended Family and Food Sharing Practices among the Contemporary Netsilik Inuit: A Case Study of Pelly Bay, NWT, Canada." (Read by Dr. G. Wenzel). 9th Inuit Studies Conference, 12–15 June, Arctic College, Nunatta Campus, Iqaluit, NWT, Canada.

1992 "Animals in the Spiritual World of Canadian Inuit: With Special Reference to Dogs in the Netsilik Inuit Society." The 7th International Abashiri Symposium "Animals in the Spiritual World of Northern Peoples", 12 November, Abashiri, Hokkaido, Japan.

1991 "Why Did the Canadian Inuit Become Christians?: A Review of the Hypotheses on the Christianization of the Canadian Inuit." International Conference on Religion and Ecology in Northern Eurasia and North America (RENENA), October 15, 1991, University of Hokkaido, Sapporo.

1987 "A Perspective on Traditional Adoptive Practices and Historical Change among Arctic Quebec Inuit of Canada." The 86th Annual Meeting of the American Anthropological Association, Chicago, November.

1987 "The Hudson's Bay Company Trade and Seasonal and Non-Seasonal Geographical Movements among the Inuit of Northern Quebec." The Annual Meeting of Canadian Ethnological Society, May, 1987, Laval University, Quebec City, Canada.

1985 "On Preferential Cousin Marriage among the Arviligyuarmiut." The 84th Annual Meeting of American Anthropological Association, December 1985, Washington D.C.

1984 "Rethinking of the Preferred First-cousin Marriage among the Traditional Netsilik Inuit." The 4th International Inuit Studies Conference, November 1984, Concordia University, Montreal, Canada, November.

1983 "Preliminary Observations on Traditional Play in Alaskan Eskimo Society." (With H. Stewart). XIth International Congress of Anthropological and Ethnological Sciences, August, 1983, Vancouver, Canada.

- Nobuhiro Kishigami has approximately 190 journal articles and book chapters, 6 single authored books, and 15 edited books in Japanese. See the following web-page.

  https://www.minpaku.ac.jp/post-staff/6566.
  https://researchmap.jp/kishigami.

References

Altman, J.C. 1987. *Hunter-gatherers today: An aboriginal economy in North Australia*. Canberra: Australian Institute of Aboriginal Studies.

Altman, J., and N. Peterson. 1988. Rights to game and rights to cash among contemporary Australian hunter-gatherers. In *Hunter-gatherers, property, power and ideology*, vol. 2., ed. T. Ingold, D. Riches, and J. Woodburn, 75–94. Oxford: Berg.

Alvard, M.S. 2002. Carcass ownership and meat distribution by big-game cooperative hunters. *Social Dimensions in the Economic Process* 21: 99–131.

Alvard, M.S., and D.A. Nolin. 2002. Rousseau's whale hunt? Coordination among big-game hunters. *Current Anthropology* 43 (4): 533–559.

Bahuchet, S. 1985. *Les Pygmees Aka et la foret centraficaine*. Paris: SELAF.

Bahuchet, S. 1990. Food sharing among the Pygmies of central Africa. *African Studies Monographs* 11 (1): 27–53.

Balikci, A. 1970. *The Netsilik Eskimos*. Garden City, N.Y.: Natural History Press.

Barnard, A. 2002. The foraging mode of thought. In *Self- and other images of hunter-gatherers* (Senri Ethnological Studies, 60, 5–24), ed. H. Stewart, A. Barnard, and K. Omura. Osaka: National Museum of Ethnology.

Belk, R. 2010. Sharing. *Journal of Consumer Research* 36 (5): 715–734.

Bennett, J., and S. Rowley (eds.). 2004. *Uqalurait: An oral history of Nunavut*. Montreal & Kingston: McGill-Queen's University Press.

Bercovitch, F.B. 1988. Coalitions, cooperation and reproductive tactics among adult and male baboons. *Animal Behaviour* 36: 1198–1209.

Berger, T.R. 1977. *Northern frontier, northern homeland: The report of the Mackenzie Valley Pipeline Inquiry*. Ottawa: Supply and Services Canada.

Berman, E. 2020. Avoiding sharing: How people help each other get out of giving. *Current Anthropology* 61 (2): 219–239.

Betzig, L., and P. Turke. 1986. Food sharing on Ifaluk. *Current Anthropology* 27 (4): 397–400.

Biesele, M., and N. Howell. 1981. The old people given you life: Aging among! Kung Hunter-gatherers. In *Other ways of growing old*, ed. P.T. Awass and S. Harrell, 77–98. Stanford: Stanford University Press.

Bird-David, N. 1990. The giving environment: Another perspective on the economic system of gatherer-hunters. *Current Anthropology* 31 (2): 189–196.

Bird-David, N. 2019. Where have all the kin gone? On hunter-gatherers' sharing, kinship and scale. In *Towards a broader view of hunter-gatherer sharing*, ed. N. Lavi and D. Friesem, 15–24. Cambridge: McDonald Institute for Archaeological Research, University of Cambridge.

Blanchet, C., and L Rochette. 2008. *Nutrition and food consumption among the Inuit of Nunavik. Nunavik Inuit health survey 2004, Qanuippitaa? How are we?* Quebec: Institut national de

N. Kishigami, *Food Sharing in Human Societies*, Trust 4,
https://doi.org/10.1007/978-981-16-7810-3

santé publique du Québec (INSPQ) & Nunavik Regional Board of Health and Social Services (NRBHSS).

Blurton Jones, N.G. 1987. Tolerated theft, suggestions about the ecology and evolution of sharing, hoarding and scrounging. *Social Science Information* 26 (1): 31–54.

Bockstoce, J.R. 2009. *Furs and frontiers in the far north: The contest among native and foreign nations for the Bering Strait fur trade.* New Heaven and London: Yale University Press.

Bodenhorn, B. 1990. I'm not the great hunter, my wife is: Iñupiat and anthropological models of gender. *Études/Inuit/Studies* 14 (1/2): 55–74.

Bodenhorn, B. 2000. It's good to know who your relatives are but we were taught to share with everybody: Shares and sharing among Inupiaq households. In *The social economy of sharing: Resource allocation and modern hunter-gatherers* (Senri Ethnological Studies 53: 27–60), ed. G.H. Wenzel, G. Hovelsrud, and N. Kishigami. Osaka: National Museum of Ethnology.

Bodenhorn, B. (2005) Sharing costs: An exploration of personal and individual property, equalities and differentiation. In *Property and equality ritualisation, sharing, egalitarianism*, vol. 1, ed. Thomas Widlok and Wolde Gossa Tadesse, 77–104. New York: Berghahn.

Brewster, K. (ed.). 2004. *The whales, they give themselves: Conversations with Harry Brower, Sr.* Fairbanks: University of Alaska Press.

Brower, H., Jr., and T. Hepa. 1998. Subsistence hunting activities and the Inupiat Eskimo. *Cultural Survival Quarterly* 22 (3): 37–39.

Buijs, C. 1993. The disappearance of traditional meat-sharing systems among the Tinitekilaamiut of East Greenland and the Arviligjuarmiut and Iglulingmiut of Canada. In *Continuity and discontinuity in arctic cultures*, ed. C. Buijs, 108–135. Leiden: Centre of Non-Western Studies.

Burch, E., Jr. 1970. The Eskimo trading partnership in North Alaska: A study in balanced reciprocity. *Anthropological Papers of the University of Alaska* 15: 49–80.

Burch, E., Jr. 1975. *Eskimo kinsmen: Changing family relationship in northwest Alaska.* St. Paul: West Publishing Co.

Burch, E. Jr. 1980. Traditional Eskimo societies in northwest Alaska. In *Alaska native culture and history* (Senri Ethnological Studies 4: 253–304), ed. Y. Kotani, and W.B. Workman. Osaka: National Museum of Ethnology.

Burch, E. Jr. 1988. Mode of exchange in North-west Alaska. In *Hunters and gatherers, Property, power and ideology*, vol. 2, ed. T. Ingold, D. Riches, and J. Woodburn, 95–109. Oxford: Berg.

Burch, E., Jr. 1998. *The Inupiaq Eskimo nations of Northwest Alaska.* Fairbanks, Alaska: University of Alaska Press.

Burch, E., Jr. 2006. *Social life in Northwest Alaska: The structure of Inupiaq Eskimo nations.* Fairbanks: University of Alaska Press.

BurnSilver, S., J. Magdanz, R. Stotts, M. Berman, and Gary Kofinas. 2016. Are mixed economies persistent or transitional? Evidence using social networks from Arctic Alaska. *American Anthropologist* 118 (1). https://doi.org/10.1111/aman.12447.

Cashdan, E. 1985. Coping with risk: Reciprocity among the Basarwa of northern Botswana. *Man* 20: 454–474.

Celsa, Y. 2002. Échange commercial et usages monétaires non-marchands dans le cadre du programme d'aide aux chasseurs du Nunavik. *Études/Inuit/Studies* 26 (2): 175–186.

Chabot, M. 2001. De la production domestique au marché: L'économie comtemporaine des familles Inuit du Nunavik. Unpublished Ph.D. thesis, Department of Sociology, Université Laval.

Chabot, M. 2003. Economic changes, household strategies, and social relations of contemporary Nunavik Inuit. *Polar Record* 39 (208): 19–34.

Collings, P. 2011. Economic strategies, community, and food networks in Ulukhatok, Northwest Territories, Canada. *Arctic* 64 (2): 207–219.

Collings, P., M.G. Marten, T. Pearce, and A.G. Young. 2016. Country food sharing networks, household structure, and implications for understanding insecurity in Arctic Canada. *Ecology of Food and Nutrition* 55 (1): 30–49.

Collings, P., G. Wenzel, and R. Condon. 1998. Modern food sharing networks and community integration in the Central Canadian Arctic. *Arctic* 51 (4): 301–314.

Condon, R. 1991. The rise of adolescence: Social change and life stage dilemmas in the Central Canadian Arctic. *Human Organization* 49 (3): 266–279.
Condon, R.P., Collings, and G. Wenzel. 1995. Best part of life: Subsistence hunting, ethnicity, and economic adaptation among young adult Inuit males. *Arctic* 48 (1): 31–46.
Council of Canadian Academies. 2014. *Aboriginal food security in Northern Canada: An assessment of the state of knowledge*. Expert panel on the state of knowledge of food security in Northern Canada. Ottawa: Canadian Council of Academies.
Dahl, J. 1989. The Integrative and cultural role of hunting and subsistence in Greenland. *Études/Inuit/Studies* 13 (1): 23–42.
Dahl, J. 2000. *Saqqaq: An Inuit hunting community in the modern world*. Toronto: University of Toronto Press.
Damas, D. 1963. *Iglulingmiut kinship and local groupings: A structural approach* (Bulletin of the National Museum of Canada, 196, Anthropological Series No.64). Ottawa: Department of Northern Affairs and National Resources.
Damas, D. 1972a. Central Eskimo systems of food sharing. *Ethnology* 11: 220–240.
Damas, D. 1972b. The structure of Central Eskimo Associations. In *Alliance in Eskimo society* (Proceedings of the American Ethnological Society, 1971, Supplement: 44–55), ed. L. Guemple. Seattle: University of Washington Press.
Damas, D. 2002. *Arctic migrants/arctic villages: The transformation of Inuit settlement in the central Arctic*. Montreal and Kingston: McGill-Queen's University Press.
de Waal, F.B.M. 1989. *Peacemaking among primates*. Cambridge, MA.: Harvard University Press.
Dombrowski, K., E. Channell, B. Khan, J. Moses, and E. Misshula. 2013. Out on the land: Income, subsistence activities, and food sharing networks in Nain, Labrador. *Journal of Anthropology* 2013 (1):https://doi.org/10.1155/2013/185048.
Donovan, G.P. 1982. *The International Whaling Commission and aboriginal/subsistence whaling: April 1979 to July 1981*. Reports of the International Whaling Commission (special Issue 4: 79–86). Cambridge: International Whaling Commission.
Dorais, L.-J. 1997. *Quaqtaq: Modernity and identity in an Inuit community*. Toronto: University of Toronto Press.
Dowling, J. 1968. Individual ownership and the sharing of game in hunting societies. *American Anthropologist* 70 (3): 502–507.
Dunn, L., and P. Gross. 2016. Food-sharing practices online in the Facebook group Cambridge Bay news. *Études/Inuit/Studies* 40 (2): 225–243.
Dyble, M. et al. 2016. Networks of food sharing reveal the functional significance of multilevel sociality in two hunter-gatherer groups. *Current Biology* 26: 2017–2021. https://doi.org/10.1016/j.cub.2016.05.064.
Ellanna, L.J., and G.K. Sherrod. 1984. *The role of kinship linkages in subsistence production* (Technical paper 100). Division of Subsistence, Alaska Department of Fish and Game.
Feit, H. 1982. The future of hunters within nation-states: Anthropology and the James Bay Cree. In *Politics and history in band societies*, ed. E. Leacock and R. Lee, 373–411. Cambridge: Cambridge University Press.
Feit, H. 1991. Gifts of the land: Hunting territories, guaranteed incomes and construction of social relations in James Bay Cree Society. In *Cash, commoditisation and changing foragers* (Senri Ethnological Studies 30: 223–268), ed. N. Peterson and T. Matsuyama. Osaka: National Museum of Ethnology.
Fienup-Riordan, A. 1983. *The Nelson Island Eskimo*. Anchorage: Alaska Pacific University Press.
Fienup-Riordan, A. 1999. *Yaqulget qaillun pilartat* (What the birds do): Yup'ik Eskimo understanding of geese and those who study them. *Arctic* 52: 1–22.
Foote, B.A. 1992. I. hunting, fishing and gathering: Methods and equipment. In *The Tigara Eskimos and their environment*, 21–32. Point Hope, Alaska: North Slope Borough Commission on Inupiat History, Language and Culture.
Fortier, J. 2000. Monkey's thigh is the shaman's meat: Ideologies of sharing among the Raute of Nepal. In *The social economy of sharing: Resource allocation and modern hunter-gatherers* (Senri

Ethnological Studies 53: 113–147), ed. G.W. Wenzel, G. Hovelsrud-Broda, and N. Kishigami. Osaka: National Museum of Ethnology.

Freuchen, P. 1981. *Book of the Eskimos*. New York: Fawcett.

Gamo, M. 1964. Structural principles of Alaskan Eskimo bands. *Japanese Journal of Ethnology* 28 (2): 149–180 (in Japanese).

George, J.C. 1981. Current procedure for allocating the bowhead whale, Balaena mysticetus, by the Eskimo whalers of Barrow, Alaska. In *Tissue structural studies and other investigations on the biology of endangered whales in the Beaufort Sea*, ed. T.F. Albert, 789–805. Report to the bureau of land management from the Department of Veterinary Science, University of Maryland.

George, J.C., M.L. Philo, G.M. Carroll, and T.F. Albert. 1988. Subsistence harvest of bowhead whales, Balaena mysticetus, by Alaskan Eskimo (SC/39/PS12). *Report of the International Whaling Commission* 38: 389–392.

Geertz, C. 1973. *The interpretation of cultures: Selected essays*. New York: Basic Books.

Gibson, T. 1985. The sharing of substance versus the sharing of activity among the Buid. *Man* (n.s.) 20: 391–411.

Gluckman, M. 1955. *The judicial process among the Barotse*. Manchester: Manchester University Press.

Godelier, M. 2000. *L'énigme du don*. Tokyo: Hosei University Press. (trans.: H. Yamauchi) (in Japanese).

Goldsmith, E. 1998. *The way: An ecological world-view*. Athens: University of Georgia Press.

Gombay, N. 2004. Making a living: Place and the commoditization of food in a Nunavik community. Unpublished Ph.D. thesis, Dept. of Geography, Queen's University.

Gombay, N. 2005a. The commoditization of country foods in Nunavik: A comparative assessment of its development, applications, and significance. *Arctic* 58 (2): 115–128.

Gombay, N. 2005b. Shifting identities in a shifting world: Food, place, community, and the politics of scale in an Inuit settlement. *Environment and Planning D: Society and Space* 23: 415–433

Gould, R. 1982. To have and have not: The ecology of sharing among hunter-gatherers. In *Resource managers: North American and Australian hunter-gatherers*, ed. N. Williams and E. Hunn, 69–92. Boulder: Westview Press.

Graburn, N.H.H. 1969. *Eskimos without Igloos: Social and economic development in Sugluk*. Boston: Little Brown and Company.

Graburn, N.H.H. 1971. Traditional economic institutions and the acculturation of Canadian Eskimos. In *Studies in economic anthropology*, ed. G. Dalton, 107–121. Washington, D.C.: American Anthropological Association.

Guemple, L. 1969. The Eskimo ritual sponsor: A problem in the fusion of semantic domains. *Ethnology* 8 (4): 468–483.

Gurven, M., K. Hill, and H. Kaplan. 2002. From forest to reservation: Transitions in food sharing behavior among the Ache of Paraguay. *Journal of Anthropological Research* 58 (1): 91–120.

Gurven, M. 2004. To give and to give not: The behavioral ecology of human food transfers. *Behavioral and brain sciences* on line. http://www.bbsonline.org/documents/a/00/00/12/57/bbs000.

Hamaguchi, H. 2002. *Introduction to whaling cultures*. Kyoto: Scitech (in Japanese).

Hamelin, L-E. 1979. *Canadian nordicity: It's your north, too*. (trans. W. Barr). Montreal: Harvest House.

Hamilton, W.D. 1964. The genetical evolution of social behavior (I). *Journal of Theoretical Biology* 7: 1–16.

Hawkes, K. 1992. Sharing and collective Action. In *Evolutionary ecology and human behavior*, ed. E.A. Smith and B. Winterhalder, 269–300. New York: Aldine De Gruyter.

Hewlett, Bonnie. 2020 (2013). *Listen, here is a story* (trans: Hattori S.T. Oishi, M. Toda). Yokohama: Shunjusha.

Hill, K. and H. Kaplan. 1988a. Tradeoffs in male and female reproductive strategies among the Ache: Part. 1. In *Human reproductive behavior: A Darwinian perspective*, ed. L. Betzig, M. Mulder, and P. Turke, 277–289. New York: Cambridge University Press.

Hill, K. and H. Kaplan. 1988b. Tradeoffs in male and female reproductive strategies among the Ache: Part. 2. In *Human reproductive behavior: A Darwinian perspective*, ed. L. Betzig, M. Mulder, and P. Turke, 277–289. New York: Cambridge University Press.

Hovelsrud-Broda, G. 1999. The integrative role of seals in East Greenlandic hunting village. *Arctic Anthropology* 36 (1/2): 37–50.

Hovelsrud-Broda, G. 2000. "Sharing", transfers, transactions and the concept of generalized reciprocity. In *The social economy of sharing: Resource allocation and modern hunter-gatherers* (Senri Ethnological Studies 53: 193–214), ed. G.W. Wenzel, G. Hovelsrud-Broda, and N. Kishigami, Osaka: National Museum of Ethnology.

Hunt, R. 2000. Forager food sharing economy: Transfers and exchanges. In *The social economy of sharing: Resource allocation and modern hunter-gatherers* (Senri Ethnological Studies. 53: 7–26), ed. G.W. Wenzel, G. Hovelsrud-Broda, and N. Kishigami. Osaka: National Museum of Ethnology.

Ichikawa, M. 1991. Consideration on egalitarianism in an evolutionary perspective. In *Natural history of homo sapien*, ed. J. Tanaka and M. Kakeya, 11–34. Tokyo: Heibonsha (in Japanese).

Ihobe, Y. 1996. Humans in far distant past: Stories before Homo sapience. In *Tracing human history: Introduction to natural anthropology*, K. Katayama et al., 1–32. Tokyo: Asakurashoten (in Japanese).

Ikuta, H. 2007. Iñupiat Pride: Kivġiq (Messenger Feast) on the Alaskan North Slope. *Études/Inuit/Studies* 31 (1/2): 343–364.

Imamura, H. 2000. *Communicating humans (homo communicans): Anthropology of gift and exchange*. Tokyo: Kodansha (in Japanese).

Imamura, K. 1993. Co-operation and sharing among the Central Kalahari San: Observation on the women's subsistence activities. *Journal of African studies* 42: 1–25 (in Japanese).

Imamura, K.1996. Small feast: Food sharing among Bushman hunter-gathers. In *Anthropology of natural society*, ed. J. Tanaka et al., 51–80. Kyoto: Academia Shupankai (in Japanese).

Ingold, T. 1988. Notes of the foraging mode of production. In *Hunters and Gathers. History, evolution and social change,* vol. 1, ed. T. Ingold, D. Riches, and J. Woodburn, 269–285. Oxford: Berg.

Irimoto, T. 2007. Development of northern studies. *Topics in contemporary cultural anthropology: From perspectives of northern studies*, ed. T. Irimoto and T. Yamagishi, 4–30, Kyoto: Sekaishisosha.

Ito, K., and Y. Kurita. 1984. *Gift exchange of Japanese*. Kyoto: Minervashobo (in Japanese).

Iwasaki-Goodman, M. 2011. Whaling activities of indigenous peoples. In *An anthropology of the marine environmental conservation* (Senri Ethnological Reports 97: 197–224), ed. H. Matsumoto, Osaka: National Museum of Ethnology.

IWC. 1981. Report of the special working group of the technical committee concerning management principles and development of guidelines about whaling for subsistence by aborigines. International Whaling Commission, IWC/33/14. Cambridge: IWC.

Jaeggi, A. V., and M. Gurven. 2013. A phylogenetic meta-analysis. *Proceedings of the Royal Society B* 280: 20131615. https://doi.org/10.1098/rspb.2013.1615 (Accessed February 8, 2021)

Kameda, T., M. Takezawa, R.S. Tindale, and C.M. Smith. 2002. Social sharing and risk reduction: Exploring a computational algorithm for the psychology of windfall gains. *Evolution and human behavior* 23: 11–33.

Kaplan, H., and K. Hill. 1985. Food sharing among Ache foragers: Tests of explanatory hypotheses. *Current Anthropology* 26 (2): 223–246.

Kaplan, H., and K. Hill. 1985. Hunting ability and reproductive success among male Ache foragers: Preliminary results. *Current Anthropology* 26 (1): 131–133.

Kaplan, H., K. Hill, K. Hawkes, and A. Hurtado. 1984. Food Sharing among Ache hunter-gatherers of Eastern Paraguay. *Current Anthropology* 25 (1): 113–115.

Kasper, C., and M. Borgerhoff Mulder. 2015. Who helps and why? Cooperative networks in Mpimbwe. *Current Anthropology* 56 (5): 701–732.

Kativik School Board. 2004. http://www.kativik.qc.ca (accessed on March 3, 2004).

Kelly, R.L. 1995. *The foraging spectrum: Diversity in hunter-gatherer lifeways*. Washington and London: Smithsonian Institution Press.
Kent, S. 1993. Sharing in an egalitarian Kalahari community. *Man* (N.S.) 28: 479–514.
Kishigami, N. 1992. On social relations related to food sharing in a contemporary society of Canadian Inuit). *Jinbun Ronkyu* 54 (Special Volume): 181–198 (in Japanese).
Kishigami, N. 1994. Why become christians? Hypotheses on the christianization of the Canadian Inuit. In *Circumpolar religion and ecology: Anthropology of the North*, ed. T. Irimoto and T. Yamada, 221–235. Tokyo: The University of Tokyo Press.
Kishigami, N. 1995. Extended family and food sharing practices among the contemporary Nestilik Inuit: A case study of Pelly Bay. *Journal of Hokkaido University of Education* (Social sciences 1-B) 45 (2): 1–9 (in Japanese).
Kishigami, N. 1998. *Arctic people, Inuit of Canada*. Tokyo: Koubundo (in Japanese).
Kishigami, N. 2000. Contemporary Inuit food sharing and hunter support program of Nunavik, Canada. In *The social economy of sharing: Resource allocation and modern hunter-gatherers* (Senri Ethnological Studies 53: 171–192), ed. G.W. Wenzel, G. Hovelsrud-Broda, and N. Kishigami. Osaka: National Museum of Ethnology.
Kishigami, N. 2002. A typology of food sharing practices among hunter-gatherers, with a special focus on Inuit examples. Paper read at CHAGS 9, Edinburgh, Scotland.
Kishigami, N. 2003. Notes on studies of food sharing in hunter gatherer societies. *Bulletin of National Museum of Ethnology* 27 (4): 725–752 (in Japanese).
Kishigami, N. 2003. On a typology of food sharing practices in hunter-gatherer societies: Transfer, exchange and re-distribution. *Japanese Journal of Ethnology* 68 (2): 145–164 (in Japanese).
Kishigami, N. 2004a. A new typology of food-sharing practices among hunter-gatherers, with a special focus on Inuit examples. *Journal of Anthropological Research* 60: 341–358.
Kishigami, N. 2004b. Cultural and ethnic identities of Inuit in Canada. In *Circumpolar ethnicity and identity* (Senri Ethnological Studies 66: 81–93), ed. T. Irimoto and T. Yamada. Osaka: National Museum of Ethnology.
Kishigami, N. 2004c. Contemporary Inuit food sharing. Paper read at the 5th International Congress of Arctic Social Sciences (ICASS V). University of Alaska, Fairbanks. May 22, 2004.
Kishigami, N. 2004d. Trends in native North American studies in Japan since the 1990s. *Japanese Review of Cultural Anthropology* 5: 91–121.
Kishigami, N. 2005a. Co-management of beluga whales in Nunavik (Arctic Quebec), Canada. In *Indigenous use and management of marine resources* (Senri Ethnological Studies 67: 121–144), ed. N. Kishigami and J. Savelle. Osaka: National Museum of Ethnology.
Kishigami, N. 2005b. Inuit, indigenous people of the Arctic Canada. In *Arctic* (*food cultures of the world* 20), ed. N. Kishigami, 121–159. Tokyo: Nobunkyo (in Japanese).
Kishigami, N. 2007 *Food culture and social change among Inuit of Canada*. Kyoto: Sekaishisosha (in Japanese).
Kishigami, N. 2008a. Notes on information and food-resources flow among the urban Inuit in Montreal, Canada. *Jinbun Ronkyu* (*Journal of Society of the Humanities*) 77: 235–246.
Kishigami, N. 2008b. A cultural anthropological study of subsistence activities with special focus on indigenous hunting, fishing and gathering in the Arctic regions. *Bulletin of the National Museum of Ethnology* 32 (4): 529–578 (in Japanese).
Kishigami, N. 2009a. A preliminary consideration of aboriginal subsistence whaling from the perspective of cultural security: A case from northwest Alaska, USA. *Bulletin of the National Museum of Ethnology* 33 (4): 493–550 (in Japanese).
Kishigami, N. 2009b. [Forum] Current status of geography and regional studies on humans and animals: Historical changes in relationship between Alaskan Iñupiaq and bowhead whales. *Human geography* 61 (5): 436–439 (in Japanese).
Kishigami, N. 2010. Food security in the Arctic regions of Canada: A case study of Nunavik Inuit society. In *Food and human securities*, ed. A. Ueda, 43–59. Osaka: Global Collaboration Center of Osaka University (in Japanese).

Kishigami, N. 2011b. The "*Apugauti*" feast of the Inupiat in Barrow, Alaska, USA. *Jinbun Ronkyu* 80: 97–110 (in Japanese).

Kishigami, N. 2012. Sharing and distribution of bowhead whale meat among the Inupiat in Barrow, Alaska, USA. *Bulletin of National Museum of Ethnology* 36 (2): 147–179 (In Japanese).

Kishigami, N. 2013a. What is a subsistence activity? With a special focus on beluga whale hunt by Inuit in Arctic Canada. *Jinbun Ronkyu* (Journal of Society of the Humanities) 82: 79–90.

Kishigami, N. 2013b. On sharing of bowhead whale meat and maktak in an Inupiat community of Barrow, Alaska, USA. *Bulletin of Hokkaido Museum of Northern Peoples* 22: 1–19.

Kishigami, N. 2013c. On the "Nalukataq" whaling festival in Barrow, Alaska, USA with a focus on the communal feast and sharing of whale meat during the festival. *Bulletin of the National Museum of Ethnology* 37 (3): 393–419 (in Japanese).

Kishigami, N. 2014. *Living with whales: Current situations among indigenous people of Alaska.* Kyoto: Rinsenshoten (in Japanese).

Kishigami, N. 2016 Reconsideration on the gift theory: Gift giving, sharing, redistribution, exchange. In *Reconsideration on the gift theory: Why do humans give to others?* ed. N. Kishigami, 10–39. Kyoto: Rinsenshoten (in Japanese).

Kishigami, N. 2018. Comments on ready and power's "Why wage earners hunt food sharing, social structure, and influence in an Arctic mixed economy" *Current Anthropology* 59 (1): 89–90.

Kishigami, N., and J.M. Savelle (eds.). 2005. *Indigenous use and management of marine resources.* Osaka. National Museum of Ethnology.

Kitanishi, K. 1996. Variability in the subsistence activities and distribution of food among different aged males of the Aka hunter-gatherers in Northeastern Congo. *African Studies Monographs* 17 (1): 35–57.

Kitanishi, K. 1997. Food sharing and residential groups of the Aka hunter-gatherers in northeastern Congo. *Journal of African Studies* 51: 1–28 (in Japanese).

Kitanishi, K. 1998. Food sharing among the Aka hunter-gatherers in northeastern Congo. *African Studies Monographs* 25 (Supplementary Issue): 3–32.

Kitanishi, 2000. The Aka and Baka: Food sharing among two Central Africa hunter-gatherer groups. In *The social economy of sharing: Resource allocation and modern hunter-gatherers.* (Senri Ethnological Studies 53: 149–169), ed. G.W. Wenzel, G. Hovelsrud and N. Kishigami. Osaka: National Museum of Ethnology.

Kitanishi, K. 2002. An owner as a sharer: Food sharing among the Aka hunter-gatherers. In *Co-existing world of forests and humans*, ed. M. Ichikawa and H. Sato, 61–91. Kyoto: Kyoto University Press (in Japanese).

Kitanishi, K. 2004. Food sharing and equality in hunting-gathering society: A case from Aka pygmies of northern Congo. In *Anthropological study of equality and inequality*, ed. H. Terashima, 53–91. Kyoto: Nakanishiya Shupan (in Japanese).

Kofinas, G., S.B. BurnSilver, J. Magdanz, R. Stotts, and M. Okada. 2016. Subsistence sharing networks and cooperation: Kaktovik, Wainwright, and Venetie, Alaska. *BOEM Report* 2015-023 DOI; AFES Report MP 2015-02. School of Natural Resources and Extension, University of Alaska Fairbanks.

Konma, T. 2000. *Cultural anthropology of gifts and exchange* (Kanagawa University Booklet 9). Tokyo: Ochanomizushobo (in Japanese).

Kruse, J.A. 1991. Alaska Inupiat subsistence and wage employment patterns: Understanding individual choice. *Arctic Anthropology* 50 (4): 317–326.

Kuchikura, Y. 2000. Optimal foraging strategies: The behavioral ecology of food acquisition. *Bulletin of the National Museum of Ethnology* 24 (4): 767–871 (in Japanese).

Langdon, S.J. 1991. The integration of cash and subsistence in Southwest Alaskan Yup'ik Eskimo communities. In *Cash, commoditisation and changing foragers* (Senri Ethnological Studies 30: 269–291), ed. N. Peterson and T. Matsuyama. Osaka: National Museum of Ethnology.

Langdon, S., and R. Worl. 1981. *Distribution and exchange of subsistence resources in Alaska* (Technical paper 55). Division of Subsistence, Alaska Department of Fish and Game.

Laugrand, F., and J. Van Oosten. 2014. *Hunters, predators, and prey: Inuit perceptions of animals.* New York, NY: Berghahn.

Lech, E. 1974 (1961). Rethinking anthropology (T. Aoki and K. Inoue trans.) London: Robert Cunningham and Sons Ltd. (in Japanese).

Leach, E. 1985 (1982). *Social anthropology* (N. Nagashima trans.) Oxford: Oxford University Press (in Japanese).

Leacock, L., and R. Lee (eds.). 1982. Introduction. In *Politics and history in band societies,* ed. E. Leacock and R. Lee, 1–20. Cambridge: Cambridge University Press.

Lee, Richard B. 1979. *The !Kung san: Men, women and work in a foraging society.* London: Cambridge University Press.

Lee, Richard B. 1993. *The dobe ju/'hoansi.* Fort Worth: Harcourt Brace College Publishers.

Lee, Richard B. 1999. Hunter-gatherer studies and the millennium: A look forward (and back). *Bulletin of the National Museum of Ethnology* 23 (4): 821–845.

Lee, R., and I. Devore, eds. 1968. *Man the hunter.* Chicago: Aldine Publishing Company.

Lefebvre, D. 1996. *Jobs in Nunavik.* Kuujjuaq, PQ: Employment and Training Department. Kativik Regional Government.

Lévesque, C., D. de Juriew, C. Lussier, and N. Trudeau. 2000. *Between abundance and scarcity: Food and the institution of sharing among the Inuit of the circumpolar region during the recent historical period.* Montreal: Universite du Quebec, INRS-urbanisation, culture et societe.

Lévi-Strauss, C. 1977 (1949). *Les structures élémentaires de la parenté.* (T. Mabuchi and S. Tajima trns.) Paris: P.U.F. (in Japanese)

Lewis, Jerome. 2019. Sharing pleasures to share rare things: Hunter-gatherers' dual distribution systems in Africa. In *Towards a broader view of hunter-gatherer sharing,* ed. Nao Lavi and David Friesem, 99–111. Cambridge: McDonald Institute for Archaeological Research, University of Cambridge.

Lonner, T.D. 1980. *Subsistence as an economic system in Alaska* (Technical paper 67), Division of Subsistence. Anchorage: Alaska Department of Fish and Game.

Macdonald, G. 2000. Economies and personhood: Demand sharing among the Wiradjuri of New South Wales. In *The social economy of sharing: Resource allocation and modern hunter-gatherers* (Senri Ethnological Studies 53: 87–111), ed. G.W. Wenzel, G. Hovelsrud-Broda, and N. Kishigami. Osaka: National Museum of Ethnology.

Magdanz, James S., Charles J. Utermohle, and Robert J. Wolfe. 2002. The production and distribution of wild food in Wales and Deering, Alaska (Technical paper 259) Juneau, Alaska: Division of Subsistence, Alaska Department of Fish and Game.

Malinowski, B. 1922. *Argonauts of the Western Pacific: An account of native enterprise and adventure in the Archipelagoes of Melanesian New Guinea.* London: Routledge and Kegan Paul.

Marlowe, F.W. 2010. *The Hadza: Hunter-gatherers of Tanzania.* Berkeley: University of California Press.

Marquardt, O., and R.A. Caulfield. 1996. Development of West Greenlandic markets for country foods since the 18th century. *Arctic* 49 (2): 107–119.

Marshall, L. 1961. Sharing, talking and giving: Relief of social tensions among the! Kung Bushmen. *Africa* 31: 321–349.

Marshall, L. 1976. *The !Kung of Nyae Nyae.* Cambridge: Harvard University Press.

Maruyama, J. 2016. With whom they share: Change and continuity in food sharing among San people. In *Reconsideration on the gift theory: Why do humans give to others?* ed. N. Kishigami, 184–208. Kyoto: Rinsenshoten (in Japanese).

Matsuyama, T. 1994. *Living in Eucalyptus forests.* Tokyo: NHK Publishing, Inc. (in Japanese).

Matsuzawa, T. 2018. *Evolution of mind to share.* Tokyo: Iwanamishoten (in Japanese).

Mauss, M. 2014 (1924). *Essai sur le don* (1924) (trans: T. Moriyama). Tokyo: Iwanamishoten (in Japanese).

Muroyama, Y. 1999. Altruistic behavior. In *For persons who study primatology,* ed. T. Nishida, and S. Uehara, 40–161. Kyoto: Sekaishisosha.

Murphy, R., and J. Steward. 1956. Tappers and trappers: Parallel process in acculturation. *Economic Development and Culture Change* 4: 335–355.

Myers, F. 1988. Burning the truck and holding the country: Property, time and the negotiation of identity among Pintupi. In *Hunter-gatherers*: *Property, Power and Ideolog*, vol. 2, ed. T. Ingold, D. Riches, and J. Woodburn, 52–74. Oxford: Berg.

Myers, Heather. 2000. Options for appropriate development in Nunavut communities. *Études/Inuit/Studies* 24 (1): 25–40.

Nadasdy, P.B. 2012. The gift in the animal: The ontology of hunting and huma-animal sociality (trans.: S. Kondo). In *Anthropology of persons and animals*, ed. K. Okuno, M. Yamaguchi, and S. Kondo, 291–348. Yokohama: Shunpusha (in Japanese).

Natcher, David C. 2019. Subsistence, regional planning and the cultural carrying capacity of First Nations in Alberta, Canada. *Hunter gatherer research* 3 (4): 717–736.

Nishida, T. 1992. Evolution of supporting behavior of primates. In *Evolution in a perspective of ecology*, ed. A. Shibatani, T. Nagano, and T. Yoro, 247–306, Tokyo: Tokyo University Press.

Nishida, T.and K. Hosaka. 2001. Food sharing among primates. In *Hominization* (Ecological anthropology 8: 255–304), ed. T. Nishida, Kyoto: Kyoto University Press (in Japanese).

Nolin, D.A. 2010. Food-sharing networks in Lamalera, Indonesia: Reciprocity, kinship, and distance. *Human Nature* 21: 243–268.

Nolin, D.A. 2012. Food-sharing networks in Lamalera, Indonesia: Status, sharing, and signaling. *Evolution and Human Behavior* 33 (4): 334–345. https://doi.org/10.1016/j.evolhumbehav.2011.11.003.

North Slope Borough. 2012. Barrow Snapshot. In North Slope Borough (ed.) 2010 Economic Profile and Census Report. Barrow: NSB Department of Administration and Finance

North Slope Borough School District. 2002. Whale Distribution in Barrow, Wainwright, Kaktovik and Nuiqsut. *Aġviqsiuġnikun, Whaling Standards, Barrow and Wainwright*. Barrow, Alaska: North Slope Borough School District.

Nuttall, M. 1991. Sharing and the ideology of subsistence in a Greenlandic sealing community. *Polar Record* 27 (162): 217–222.

Nuttall, M. 1992. *Arctic homeland: Kinship, community, and development in Northwest Greenland*. Toronto: University of Toronto Press.

Oda, R. 2019. Four types of exchange and their mixture. *The Journal of social sciences and humanities* 515-2 (social anthropology 12: 1–22) (in Japanese).

Omura, K. 2011. Living dually: Social positions of subsistence and production in Inuit society of Canada. In *Globalization and living world: Anthropological contemporary in a perspective of subsistence*, ed. K. Matsui, K. Nawa, and A. Nobayashi, 65–96. Tokyo: Institute for Advanced Studies on Asia.

Omura, K. 2021. The engines of world systems. *Japanese Journal of Cultural Anthropology* 86 (1): 57–75 (in Japanese).

Paine, R. (ed.). 1977a. Introduction. In *The white Arctic* (Newfoundland Social and Economic Papers No.7), ed. R. Paine. St. Johns, Newfoundland: Institute of Social and Economic Research, Memorial University of Newfoundland.

Paine, R. 1977b. The path to welfare colonialism. In *The white Arctic: Anthropological essays on tutelage and ethnicity*, ed. R. Paine, 7–28, St. John's. NF: Institute of Social and Economic Research, Memorial University of Newfoundland.

Petersen, R. 1989. Traditional and contemporary distribution channels in subsistence hunting in Greenland. In *Keynote speeches from the Sixth Inuit Studies Conference, Copenhagen, October 1988*, ed. J. Dahl. Copenhagen: Institute of Eskimology, Copenhagen University.

Peterson, N. 1993. Demand sharing: Reciprocity and the pressure for generosity among foragers. *American Anthropologist* 95: 860–876.

Peterson, N., and T. Matsuyama (eds.). 1991. *Cash, commoditisation and changing foragers* (Senri Ethnological Studies 30). Osaka: National Museum of Ethnology.

Polanyi, K. 1957. Economy as instituted process. In *Trade and Market in the early empires*, ed. K. Polanyi, 243–270. New York: The Free Press.

Polanyi, K. 1975 (1966). *Dahomey and the slave trade: An analysis of an archaic economy.* S. Kurimoto and N. Hata trans. Seattle: University of Washington Press (in Japanese).

Polanyi, K. 1977. *The livelihood of man.* In H.W. Pearson (ed.). New York: Academic Press.

Price, J. 1975. Sharing: The integration of intimate economies. *Anthropologica* (n.s.) 17(1): 3–27.

Pryor, F.L., and N.H.H. Graburn. 1980. The myth of reciprocity. In *Social exchange: Advances in theory and research*, ed. K.J. Gergen, M.S. Greenberg, and R.H. Wills, 215–237. New York and London: Plenum Press.

Quintal-Marineau, M. 2016. Near the floe edge: Inuit women's role in the mixed economy. Unpublished PhD dissertation, Department of Geography, McGill University, Montreal.

Quintal-Marineau, M. 2017. The new work regime in Nunavut: A gender perspective. *Canadian Geographer* 61 (3): 334–345.

Quintal-Marineau, M. 2019. Feeding our families; That's what we have been doing for centuries': Inuit women's sharing practices and strategies. *Hunter Gatherer Research* 3 (4): 583–599.

Quintal-Marineau, M., and G. Wenzel. 2019. Men hunt, women share: Gender and contemporary Inuit subsistence relations. In *Towards a broader view of hunter-gatherer sharing*, ed. Nao Lavi and David Friesem, 211–220. Cambridge: McDonald Institute for Archaeological Research, University of Cambridge.

Rainey, F. 1947. The whale hunters of Tigara. *Anthropological Papers of the American Museum of Natural History* 41 (2).

Riches, D. 1981. The obligation to give: An interactional sketch. In *The Structure of folk models* (Monograph 20: 209-203) ed. L. Holy, and M. Stuchlik. London: Academic Press.

Ready, E. 2016. Food, sharing, and social structure in an Arctic mixed economy. PhD. dissertation submitted to the Department of Anthropology and the Committee on Graduate Studies of Stanford University, USA.

Ready, E. 2018. Sharing-based social capital associated with harvest production and wealth in the Canadian Arctic. *PloS One* 13 (3): e0193759.

Ready, E. 2019. Why subsistence matters. *Hunter Gatherer Research* 3 (4): 635–650.

Ready, E., and E.A. Power. 2018. Why wage earners hunt: Food sharing, social structure, and influence in an Arctic mixed economy. *Current Anthropology* 59 (1): 74–97.

Robbe, P. 1994. *Les Inuit d'Ammassalik, Chasseurs de l'Arctique (Mémoires du Muséum National d'Histoire Naturelle Tome 159, Ethnogogie).* Paris: Muséum National d'Histoire Naturelle.

Sahlins, M. 1965. On the sociology of primitive exchange. In *The Relevance of models for social anthropology* (Monographs of the Association of Social Anthropologists, No.1: 139–236), ed. M. Banton, London: Tavistock.

Sahlins, M. 1984 (1972). *Stone age of economics.* H. Yamauchi, trans Chicago: Aldine Publishing Company (in Japanese).

Sakakibara, C. 2010. Kiavallakkikput aġviq (Into the whaling cycle): Centaceousness and climate change among the Iñupiat of Arctic Alaska. *Annals of the Association of American Geographers* 100 (4): 1003–1012.

Savelle, J.M. 2005. The development of indigenous whaling: Prehistoric and historic contexts. In *Indigenous use and management of marine resources* (Senri Ethnological Studies 67: 53–58), ed. N. Kishigami, and J.M. Savelle.Osaka: National Museum of Ethnology.

Scott, C. 1984. Between "Original affluence" and consumer affluence: Domestic production and guaranteed income for James Bay Cree hunters. In *Affluence and cultural survival* (1981 Proceedings of the American Ethnological Society), ed. R. F. Salisbury, and E. Tooker, 74–86. Washington, D.C.: the American Ethnological Society

Searles, E. 2002. Food and the making of modern Inuit identities. *Food and Foodways* 10: 55–78.

Service, E. 1966. *The hunters.* Englewood Cliffs, N.J.: Prentice-Hall.

Service, E. 1987. Social behavior in evolutionary perspective. In *Primate Societies*, B.B. Smuts et al., 319–329. Chicago: The University of Chicago Press.

Sheehan, G. 1997. *In the belly of the whale: Trade and war in Eskimo society.* Anchorage: Alaska Anthropological Association.

Sillander, Kenneth. 2019. The demand for closeness: Social incentives for sharing among hunter-gatherers and other groups. In *Towards a broader view of hunter-gatherer sharing*, ed. N. Lavi and D. Friesem, 71–83. Cambridge: McDonald Institute for Archaeological Research, University of Cambridge.
Silk, J.B. 1987. Social behavior in evolutionary perspective. In *Primate societies*, B.B. Smuts et al., 319–329. Chicago: The University of Chicago Press.
Smith, B.C., and J.C. Brigham. 1992. Native radio broadcasting in North America: An overview of systems in the United States and Canada. *Journal of Broadcasting and Electric Media* 36 (2): 183–194.
Smith, E A. 1988. Risk and uncertainty in the 'Original affluent society': Evolutionary ecology of resource-sharing and land tenure. In *Hunters and Gatherers, history, evolution and social change*, vol. 1, ed. T. Ingold, D. Riches, and J. Woodburn, 222–251. Oxford: Berg.
Smith, E.A., and R. Boyd. 1990. Risk and Reciprocity: Hunter-gatherer socioecology and the problem of collective action. In *Risk and Uncertainty in tribal and peasant economics*, ed. E. Cashdan, 167–192. Boulder: Westview Press.
Smith, E.A., and B. Winterhalder, eds. 1992. *Evolutionary Ecology and human behavior*. New York: Aldine De Gruyter.
Smith, E., et al. 2003. The benefits of costly signaling: Meriam turtle hunters. *Behavioral Ecology* 14 (1): 116–126.
Sonoda, K. 2019. Language socialization of inappropriate words: *Bangà* and Baka hunter-gatherer societies' children. *Japanese journal of cultural anthropology* 84 (3): 243–261 (in Japanese).
Spencer, R. 1959. *The North Alaskan Eskimo: A study in ecology and society*. Washington, D.C.: Bureau of American Ethnology.
Stairs, A., and G. Wenzel. 1992. I am I and the environment: Inuit hunting, community, and identity. *The Journal of Indigenous Studies* 3 (1): 1–12.
Statistics Canada. 2005. *2001 Census Dictionary* (Internet Version). Catalogue No. 92-378-XIE.
Stephen R. Braund & Associates. 2018. Description of Alaska Eskimo bowhead subsistence sharing practices: Including an overview of bowhead whale harvesting and community-based need (final report). A report to submitted to Alaska Eskimo whaling commission, Utqiaġvik, Alaska, USA. http://www.north-slope.org/assets/images/uploads/Braund_AEWC16_Bowhead_Sharing_Report_5-23-18.pdf (accessed August 15, 2020).
Stewart, H. 1991. Gender roles in food sharing. *Annual Report of Social Anthropology* 17: 115–127 (in Japanese).
Suydam, R. et al. 2006. Subsistence Harvest of Bowhead Whales (Balaena mysticetus) by Alaskan Eskimos during 2005. Barrow: Alaska Whaling Commission.
Suydam, R. et al. 2007. Subsistence Harvest of Bowhead Whales (Balaena mysticetus) by Alaskan Eskimos during 2006. Barrow: Alaska Whaling Commission.
Suydam, R. et al. 2008. Subsistence Harvest of Bowhead Whales (Balaena mysticetus) by Alaskan Eskimos during 2007. Barrow: Alaska Whaling Commission.
Suydam, R. et al. 2009. Subsistence Harvest of Bowhead Whales (Balaena mysticetus) by Alaskan Eskimos during 2008. Barrow: Alaska Whaling Commission.
Suydam, R. et al. 2010. Subsistence Harvest of Bowhead Whales (Balaena mysticetus) by Alaskan Eskimos during 2009. Barrow: Alaska Whaling Commission.
Suydam, R. et al. 2011. Subsistence Harvest of Bowhead Whales (Balaena mysticetus) by Alaskan Eskimos during 2010. Barrow: Alaska Whaling Commission.
Takekawa. 2008. Taboos and resources: How do humans explain nature? In *Anthropology of distribution and management of marine resources*, ed. N. Kishigami, 245–270. Tokyo: Akashishoten.
Takeuchi, K. 1995. Ritual aspects and pleasure in hunting activity: Cooperation and distribution in the net-hunting activity of Aka hunter-gatherers in northern Congo. *African studies* 46: 57–76 (in Japanese).
Tanaka, J. 1980. *The san, hunter-gatherer of the Kalahari: A study in ecological anthropology*. Tokyo: University of Tokyo Press.

Tanno, Tadashi. 1991. Sharing as sharing life. In *Natural history of Homo sapiens*, J. Tanaka, and M. Kakeya, 11–34, Tokyo: Heibonsha (in Japanese).

Terashima, H. (ed.) 2021. *Sharing/storing* (Session 2: challenges of ecological anthropology). Kyoto: Kyoto University Press (in Japanese).

Testart, A. 1987. Game sharing systems and kinship systems among hunter-gatherers. *Man* (N.S.) 22: 287–304.

Trivers, R.L. 1971. The Evolution of reciprocal altruism. *Quarterly Review of Biology* 46: 35–57.

Tucker, B. 2004. Giving, scrounging, hiding, and selling: Minimal food transfers among Mikea forager-farmers of Madagascar. *Research in Economic Anthropology* 23: 43–66.

Tucker, B. 2019. Ecological, historical and social explanations for low rates of food sharing among Mikea foragers of southwest Madagascar. In *Towards a broader view of hunter-gatherer sharing*, ed. N. Lavi and D. Friesem, 237–250. Cambridge: McDonald Institute for Archaeological Research, University of Cambridge.

Turner, E. 1990. The whale decides: Eskimos' and ethnographer's shared consciousness on the ice. *Études/Inuit/studies* 14 (1/2): 39–52.

Van de Velde, F. 1956. Rules for sharing the seal amongst the Arviligjuarmiut Eskimo. *Eskimo* 41: 3–6.

VanStone, J.W. 1962. *Point Hope: An Eskimo village in transition*. Seattle: University of Washington press.

Victor, Anne-Marie. 1987. Éléments symboliques de la chasse à la baleine. *Études/Inuit/Studies* 11 (2): 139–163.

Wenzel, G.W. 1981. Clyde Inuit adaptation and ecology: The organization of subsistence. Canadian Ethnology Service Mercury Paper No.77. Ottawa: National Museum of Man.

Wenzel, G.W. 1991. *Animal rights, human rights: Ecology, economy and ideology in the Canadian Arctic*. Toronto: University of Toronto Press.

Wenzel, G.W. 1995. Ningiqtuq: Resource sharing and generalized reciprocity in Clyde River, Nunavut. *Arctic Anthropology* 32 (2): 43–60.

Wenzel, G.W. 1996. Inuit sealing and subsistence managing after the E.U. sealskin Ban. *Geographische Zeitschrift* 84 (3/4): 130–142.

Wenzel, G.W. 2000. Sharing, money, and modern Inuit subsistence: Obligation and reciprocity at Clyde River, Nunavut. In *The social economy of sharing resource allocation and modern hunter-gatherers* (Senri Ethnological Studies 53: 61–85), ed. G.W. Wenzel, G. Hovelsrud Broda, and N. Kishigami, Osaka: National Museum of Ethnology.

Wenzel, G.W. 2013. Inuit and modern hunter-gatherer subsistence. *Études/Inuit/Studies* 37(2): 181–200.

Wenzel, G.W. 2019. Canadian Inuit subsistence. *Hunter Gatherer Research* 3 (4): 567–581.

Wenzel, G.W., G. Hovelsrud-Broda, and N. Kishigami (eds.). 2000. *The social economy of sharing: Resource allocation and modern hunter-gatherers* (Senri Ethnological Studies 53: 1–6) Osaka: National Museum of Ethnology.

Widlok, T. 2013. Sharing: Allowing others to take what is valued. *HAU: Journal of Ethnographic Theory* 3(2): 11–31.

Widlok, T. 2017. *Anthropology and the economy of sharing*. London and New York: Routledge.

Widlok, T. 2019. Extending and limiting selves: A processual theory of sharing. In *Towards a broader view of hunter-gatherer sharing*, ed. Nao Lavi and David Friesem, 25–38. Cambridge: McDonald Institute for Archaeological Research, University of Cambridge.

Wiessner, P. 1982. Risk, reciprocity, and social influence on! Kung San economies. In *Politics and history in band societies*, ed. E. Leacock and R. Lee, 61–84. Cambridge: Cambridge University Press.

Wiessner, P. 1986. !Kung San networks in a generational perspective. In *The past and future of !Kung ethnography: Critical reflections an symbolic perspectives, essays in honour of Loma Marshall*, ed. M. Biesele, R. Gordon, and R. Lee, 103–136. Hamburg: Helmut Buske Verlag.

Wiessner, P. 1998. On network analysis: The potential for understanding (and misunderstanding) !Kung Hxaro. *Current Anthropology* 39 (4): 514–519.

Winterhalder, B. 1986. Diet choice, risk, and food sharing in a stochastic environment. *Journal of Anthropological Archaeology* 5: 369–392.

Winterhalder, B. 1986. Optimal foraging: Simulation studies of diet choice in a stochastic environment. *Journal of Ethnobiology* 6: 205–223.

Winterhalder, B., and E.A. Smith. 1981. *Hunter-gatherer foraging strategies: Ethnographic and Archaeological analyses*. Chicago and London: The University of Chicago Press.

Woodburn, J. 1982. Egalitarian societies. *Man* (N.S.) 17: 431–451.

Woodburn, J. 1998. Sharing is not a form of exchange: An analysis of property-sharing in immediate-return hunter-gatherer societies. In *Property relations: Renewing the anthropological tradition*, ed. C.M. Hann, 48–63. Cambridge: Cambridge University Press.

Worl, R. 1980. The North Slope Inupiat whaling complex. In *Alaska native culture and history* (Senri Ethnological Studies 4: 305–320), ed. Y. Kotani, and W.B. Workman, Osaka: National Museum of Ethnology.

Worl, R. and C. W. Smythe. 1986. *Barrow: A decade of modernization: The Barrow case study*. Prepared for minerals management service, Alaska OCS region, Alaska OCS socioeconomic studies program. U.S. department of the interior technical report No. 125.

Yamagiwa, J. 1997. When did homo sapience become human beings? In *New discovery of human beings*, ed. T. Aoki et al., 31–60. Tokyo: Iwanamishoten (in Japanese).

Ziker, J.P. 1998. Kinship and exchange among the Dolgan and Nganasan of northern Siberia. *Research in Economic Anthropology* 19: 191–238.

Ziker, J.P. 2007. Subsistence and food sharing in northern Siberia: Social and nutritional ecology of the Dolgan and the Nganasan. *Ecology of Food and Nutrition* 46 (5/6): 445–467. https://doi.org/10.1080/03670240701486743.

Ziker, J.P., and M. Schnegg. 2005. Food sharing at meals: Kinship, reciprocity, and clustering in the Taimyr Autonomous Okrug, northern Russia. *Human Nature* 16 (2): 178–210.

Ziker, J. P., Karen S. Fulk. 2018. Paying it forward or giving back? Women's sharing networks in Siberia. *Cross-Cultural Research* 53 (3): 272–290. https://doi.org/10.1177/1069397118806821.

Index

A
Aboriginal Subsistence Whaling, 71, 90
Aché, 19–21, 28, 99, 100, 102, 106–109, 111, 116–118, 120–123, 125
Africa, 16, 23, 99, 100, 102, 123, 128–130
Aka, 10–12, 15–17, 21, 38, 99, 100, 102, 103, 105, 106, 116–118, 120–122, 125, 132
Akulivik, 14, 40, 41, 43–57, 60–65, 112–115
Alaska, 13, 23, 25, 26, 29, 30, 37, 40, 67–69, 71–73, 75, 76, 83–85, 97, 98, 123, 127, 132, 134
Altman, J., 2, 12, 64, 123
Amount, 3, 6, 7, 12, 15, 17, 19, 21, 22, 26, 28, 32, 33, 36, 37, 47, 48, 54, 55, 57, 63, 64, 70, 82–84, 88–90, 92, 97, 101–106, 108–111, 113, 115, 117–120, 122, 130, 131
Arctic Canada, 23, 26, 30, 31, 112
Australian Aboriginal, 2, 10–12, 14, 38, 121, 123, 128, 131

B
Balanced reciprocity, 4, 5, 20, 33, 129
Balikci, A., 30, 31, 123
Barnard, A., 1, 10, 15
Barrow, 29, 40, 67, 73, 75, 76, 79, 83
Behavioral ecology, 1, 18, 101
Belk, R., 1, 2, 122–124, 126
Beluga whale, 30, 33, 35, 36, 45, 47–50, 53–56, 61, 65, 67, 85, 91, 94–96, 112, 114, 120, 123, 127, 132
Bird-David, N., 13, 14, 38, 102, 120, 126, 127, 133
Blurton Jones, N., 20, 21
Bodenhorn, B., 13, 14, 29, 39, 40, 63, 86, 90–96, 98, 102, 118, 120, 123, 127, 128, 133, 134
Bowhead whale, 21, 25, 26, 30, 67, 68, 70, 74, 75, 79–81, 85–87, 91, 95, 96, 98, 112, 120, 123, 127, 128, 132, 133
Burch, Jr. E., 12, 29, 39, 40, 68–70, 73, 123, 127

C
Camp, 15–17, 19–21, 31–34, 38–40, 43, 46–48, 64, 99, 102–113, 115–117, 119, 120, 122, 125, 126, 130, 132
Caribou, 30, 33, 45, 46, 54–58, 66, 67, 69, 70, 84, 85, 95, 96, 112–115, 121, 127
Christianity, 43, 93
Collings, P., 23, 24, 30, 33, 64, 65, 133
Comparison, 37, 65, 73, 79, 81, 100, 108, 122, 123
Condon, R., 24, 30, 33, 65, 133
Cooked food, 17, 103, 105, 108, 116, 117, 121
Copper Inuit, 31, 33
Costly signaling, 18, 22, 28
Cree, 43, 44, 66

D
Dahl, J., 2, 3, 35, 36, 91, 96
Damas, D., 1, 12, 30–32, 35, 39, 64, 113, 123, 127, 129
Definition (of sharing), 1, 2, 121, 123
Demand sharing, 6, 8, 20, 21, 33, 96, 102, 114, 121, 123, 128
Dolgan, 24, 127

N. Kishigami, *Food Sharing in Human Societies*, Trust 4,
https://doi.org/10.1007/978-981-16-7810-3

Dorais, L-J., 34, 127

E
Ecological adaption, 15, 16, 27, 28, 134
Environmental view, 102, 119–121, 126, 128, 129, 133
European Community (EC), 44
Evolutionary ecology, 1, 18, 96, 101, 108
Exchange, 1–9, 12, 13, 15, 20–24, 26, 28, 29, 31, 35–37, 61, 62, 64, 69, 81, 85, 86, 89–91, 95, 98–100, 102, 103, 116, 122, 124, 126, 127, 129, 130, 134–136
Extended family, 24, 25, 29, 31–34, 37, 44, 46–48, 50–52, 61–65, 68–70, 84, 89, 92, 113–115, 117, 118, 120, 127–130, 132, 133

F
Family, 12, 15, 19, 20, 24–26, 28–30, 32, 34, 35, 39, 44, 46, 51, 68–70, 72, 79, 84–86, 92, 94, 95, 97, 102–111, 113–120, 122, 125, 127, 129, 130, 132, 136
Farmer, 102, 103
Fienup-Riordan, A., 13, 29, 30, 40, 63, 98, 102, 113, 118, 120, 133, 134
FM radio, 47, 49, 50, 53, 61
Food sharing through meals, 51
Form, 1–4, 6, 7, 10, 11, 13, 25, 29, 31, 32, 34, 35, 37, 39, 46, 47, 49, 62, 70, 93, 95, 98–100, 102, 108, 113, 115–117, 121, 122, 124, 126, 128, 129, 132–136
Freuchen, P., 124, 130

G
Generalized exchange, 4, 5
Generalized reciprocity, 4, 5, 29, 33, 36, 70, 129
George, C., 73, 79–81, 86
Gift, 3, 4, 6, 8, 12, 15, 30, 36, 50, 51, 82, 84, 91, 122–124, 126, 129, 130
Gift-giving, 2–4, 6–8, 12, 13, 123, 126, 129, 130, 136
Giving, 3, 6–8, 12, 13, 20, 22, 24, 26, 29–31, 33, 34, 36, 37, 39, 44, 47, 50, 62, 63, 77, 89–91, 99–101, 116–119, 121, 122, 124, 126–130, 133, 135, 136
Graburn, N., 2, 3, 20, 33, 34, 47, 96, 102, 123, 129
Greenland, 3, 14, 29, 35–37, 40, 91, 98, 124
Gurven, M., 18, 22, 23, 107, 109, 110, 116, 120, 123, 134

H
Hadza, 28, 123, 129
Hill, K., 18–21, 28, 102, 107–110, 120, 123
Honey, 15, 19, 28, 103, 105, 107–111
Household, 14, 23–26, 32–37, 46–48, 50–57, 62–65, 68, 69, 79, 82, 84, 89, 92, 95, 98, 105, 107, 112–115, 117, 129, 130, 136
Hovelsrud-Broda, G., 14, 29, 35, 40, 63, 96
Hunter support program, 34, 44, 47, 53–58, 60, 62–66, 113–116, 132
Hunt, R., 2, 7
Hxaro, 15, 23

I
Ichikawa, M., 3, 12, 14
Identity, 12, 63, 64, 67, 92, 96, 101, 116, 133
Igloolik Inuit, 31, 32
Imamura, H., 12, 13
Imamura, K., 3, 11, 16, 17, 21, 38, 39, 99, 123, 125
Indebtedness, 101, 118, 119, 122, 124, 130
Ingold, T., 17, 39, 99, 125
Insurance, 12, 15, 28, 32, 37, 63, 65, 101, 131
International Whaling Commission (IWC), 71, 90
Inuit, 1–3, 10–12, 14, 18, 21, 24–27, 29–41, 43–49, 53–66, 90, 96, 97, 99, 100, 102, 112–122, 124–130, 132–135
Iñupiaq, *see* Iñupiat
Iñupiat, 11, 21, 29, 30, 37–40, 120, 123

J
James Bay and Northern Quebec Agreement (JBNQA), 34, 44, 53

K
Kalahari, 12, 16
Kaplan, H., 18–21, 28, 102, 107–111, 116, 120, 123
Kativik regional government, 54, 114

Kin, 6, 18–20, 24, 28, 47, 53, 68, 92, 95, 101, 111, 113, 118, 122, 123
Kinship, 4, 5, 10–13, 16, 17, 19, 24, 25, 30, 31, 38–40, 44, 48, 63–65, 103, 105, 106, 108, 110, 111, 113, 115, 120, 126–129
Kinsmen, 47–49, 51–53, 55, 58, 62, 65
Kishigami, N., 1, 7, 12, 14, 21, 22, 25, 29–31, 34, 39, 40, 43, 47, 49–51, 54, 55, 57, 62, 63, 72, 73, 75, 76, 82, 83, 93, 96–102, 112–114, 116, 118–123, 125–128, 132, 133, 136
Kitanishi, K., 3, 11, 14–17, 21, 38, 39, 102–106, 120
Kofinas, G., 23, 25, 84, 85
Konma, T., 2, 6, 23, 134
Kula, 3

L
Leach, E., 4, 7
Levelling, 12, 15, 28, 101, 106, 115, 117, 122, 132
Lévi-Strauss, C., 3–6

M
Malinowski, B. K., 3
Māori, 3
Matsuzawa, T., 23, 134, 136
Mauss, M., 3, 4, 12, 101, 118, 122, 123, 129, 130, 135, 136
Mbuti, 12, 120
Mikea, 21, 26
Missionary, 31, 43, 107, 112
Mixed economy, 25, 26, 45, 112
Myers, F., 38, 130

N
Nalukataq, 74, 76–80, 82, 83, 88, 89, 92, 132
Nayaka, 13, 14, 120, 128, 133
Negative reciprocity, 4, 5, 129
Netsilik Inuit, 13, 30, 31, 123
Nganasan, 24, 127
Nolin, D., 23, 24, 96
North America, 1, 3, 8, 12, 13, 26, 39, 41, 98–100, 121, 126, 132
Nunavik, 14, 24, 26, 34, 40, 41, 43–48, 53, 60, 64, 65, 113, 114, 123, 132
Nuttall, M., 13, 21, 35, 36, 40, 63, 102, 113, 120, 127

P
Paine, R., 43
Peterson, N., 2, 6–8, 12, 14, 20, 21, 100, 102, 121, 123, 128, 131
Plant food, 19, 22, 104, 106, 107
Point Hope, 71, 72, 84, 85, 97
Polanyi, K., 5, 7
Political power, 26, 61, 63, 101, 118, 132
Population, 19, 25, 26, 32, 38–40, 43, 46, 54, 57, 65, 67, 84, 98, 102, 105, 111, 112, 114, 117–120, 122, 126, 130–132
Price, J., 1, 6, 123, 125
Primary sharing, 16, 17, 81, 90, 96, 103, 104, 106, 113, 116, 121
Primate, 1, 2, 22, 23, 28, 134, 135
Property, 4–6, 12, 20, 68, 69, 94
Pygmy, 11, 102, 103

Q
Quebec, 34, 40, 41, 43–45, 53, 54, 113, 114

R
Ready, E., 12, 22–26, 39, 61, 113, 127, 129, 132, 133
Reciprocal altruism, 18, 20, 24, 28, 101, 123
Reciprocity, 1, 3–8, 19, 20, 28–32, 35, 36, 38, 70, 89, 91, 95, 127, 129
Redistribution, 5, 7–9, 25, 30, 32, 37, 54, 57, 64, 69, 99, 100, 115–117, 122, 124, 126, 129
Reputation, 7, 22, 35, 37, 61, 63, 93, 96, 101, 118, 119, 122, 134
Riches, D., 34, 39, 62
Risk, 12, 15, 28, 101, 117, 122, 131

S
Sahlins, M., 2–5, 20, 32, 129
San, 10–12, 14–17, 22, 38, 123
Seal, 13, 30–33, 35, 36, 43–46, 49, 50, 54, 56–58, 61, 67, 69, 70, 82, 85, 94–97, 112, 113, 123
Searles, N., 33
Secondary sharing, 11, 16, 89, 103–106, 116
Sedentarization, 14, 18, 40, 43, 64
Settlement, 12, 14, 18, 19, 35, 36, 38, 43, 99, 125–128, 130–132, 134
Show off, 63, 101
Smith, E., 15, 18, 19, 22, 29, 38, 119, 130

Social change, 14, 30, 64, 65
Social network, 23–26, 29, 38
Social relation, 3, 4, 10–12, 18, 21–23, 26–30, 34–36, 39, 63, 91, 129, 131
Social relationship, 44, 52, 53, 62–65, 92, 96, 101, 106, 111, 115, 117, 118, 127, 133, 135, 136
South America, 99, 100, 106, 128, 129
Subsistence, 11, 13, 24, 26, 27, 30, 31, 33, 36, 37, 39, 44–46, 57, 63, 64, 67, 71, 93, 98, 102, 106, 112, 125, 134
Superiority-inferiority relation, 101, 119

T
Tanno, T., 3, 11, 17, 39, 99
Tertiary sharing, 3, 16, 17, 81, 103, 104, 106, 116
Testart, A., 10, 11
Tolerated theft, 2, 18, 20, 21, 28, 101, 123, 134
Transfer, 2, 3, 5–9, 35, 43, 46, 48, 58, 62, 64, 68, 69, 99, 100, 116, 118, 122–124, 126, 129, 133
Tucker, B., 18, 21, 26, 27

U
Utqiaġvik, 29, 40, 67, 68, 71–73, 75, 76, 79, 81, 83–87, 89, 91–94, 96–98

W
We feelings, 63, 116
Welfare colonialism, 43
Well-being, 12, 24, 25, 36, 37, 94, 96, 97, 118, 132, 134
Wenzel, G., 12, 14, 21, 24, 27, 29, 30, 32, 33, 39, 40, 44, 45, 63, 65, 96, 97, 102, 113, 118, 120, 127, 128, 133, 134
Whaling captain, 71, 73, 74, 79, 84, 86, 89, 92, 93, 96, 97, 128, 132
Widlok, T., 1, 2, 12, 22, 23, 28, 99, 100, 102, 121, 123, 125–132, 134–136
Wiessner, P., 15, 23
Winterhalder, B., 15, 18, 19, 38, 119, 130
Woodburn, J., 1, 7, 9, 10, 12, 64, 99, 100, 123, 127, 129, 130, 132
World view(s), 13, 14, 28, 30, 35, 37

Y
Yupik, *see* Yupiit

Z
Ziker, J., 12, 18, 19, 23, 24, 127